DEPARTMENT OF TRANSPORT

International Code of Signals 1969

LONDON: HMSO

© Crown copyright 1969
First published 1969
Consolidated Edition 1991

ISBN 0 11 551015 X

Preface

Codes of signals for the use of mariners have been published in various countries since the beginning of the nineteenth century.

The first International Code was drafted in 1855 by a committee set up by the British Board of Trade. It contained 70,000 signals, it used eighteen flags and it was published by the British Board of Trade in 1857 in two parts; the first containing universal and international signals and the second British signals only. The book was adopted by most seafaring nations.

This edition was revised by a committee set up in 1887 by the British Board of Trade. This committee's proposals were discussed by the principal maritime powers and at the international conference in Washington in 1889. As a result, many changes were made, the Code was completed in 1897 and was distributed to all maritime powers. This edition of the **International Code of Signals,** however, did not stand the test of World War 1.

The International Radiotelegraph Conference at Washington in 1927 considered proposals for a fresh revision of the Code and decided that it should be prepared in seven editorial languages, namely in English, French, Italian, German, Japanese, Spanish and in one Scandinavian language which was chosen by the Scandinavian governments to be the Norwegian language. The new edition was completed in 1930 and was adopted by the International Radiotelegraph Conference held in Madrid in 1932. The new Code was compiled in two volumes, one for use by visual signalling and the other by radiotelegraphy. Words and phrases applicable to aircraft were introduced in volume II together with a complete Medical Section and a code for accelerating the granting of pratique. The Medical Section and the pratique signals were prepared with the assistance and by the advice of the Office International d'Hygiène Publique. The Code, particularly volume II, was primarily intended for use by ships and aircraft and, via coast radio stations, between ships or aircraft and authorities ashore. A certain number of signals were inserted for communications with ship-owners, agents, repair yards, etc. The same conference (Madrid, 1932) established a standing committee to review the Code, if and when necessary, to give guidance on questions of use and procedure and to consider proposals for modifications. Secretarial duties were undertaken by the government of the United Kingdom. The standing committee met only once in 1933 and introduced certain additions and amendments.

The Administrative Radio Conference of the International Telecommunication Union suggested in 1947 that the **International Code of Signals** should fall within the competence of the Inter-Governmental Maritime Consultative Organization (IMCO). In January 1959, the first Assembly of IMCO decided that the Organization should assume all the functions then being performed by the Standing Committee of the International Code of Signals. The second Assembly in 1961 endorsed plans for a comprehensive review of the **International Code of Signals** intended to meet the present-day requirements of mariners. A sub-committee of the Maritime Safety Committee of the Organization was established to revise the Code, to prepare it in nine editorial languages, namely the original seven (English, French, Italian, German, Japanese, Spanish and Norwegian) together with Russian and Greek, and to consider proposals for a new radiotelephone code and its relation to the **International Code of Signals**. The sub-committee consisted of representatives of the following countries: Argentina, Federal Republic of Germany, France, Greece, Italy, Japan, Norway, Union of Soviet Socialist Republics, United Kingdom and the United States of America. The following international governmental and non-governmental organizations contributed to, and assisted in, the preparation of the revised Code: the International Atomic Energy Agency, the International Civil Aviation Organization, the International Labour Organization, the International Telecommunication Union, the World Meteorological Organization, the World Health Organization, the International Chamber of Shipping, the International Confederation of Free Trade Unions and the International Radio-Maritime Committee.

The sub-committee completed the revision of the Code in 1964, taking into account Recommendation 42 of the 1960 Conference on Safety of Life at Sea and Recommendation 22 of the Administrative Radio Conference, Geneva 1959.

The revised Code is intended to cater primarily for situations related essentially to safety of navigation and persons, especially when language difficulties arise. It is suitable for transmission by all means of communication, including radiotelephony and radiotelegraphy, thus obviating the necessity for a separate radiotelephone code and dispensing with volume II for radiotelegraphy. The revised Code embodies the principle that each signal has a complete meaning. It thus leaves out the vocabulary method which was part of the old Code. The Geographical Section, not being considered essential, was omitted. By these means it was possible to reduce considerably the volume of the Code and achieve simplicity. Suitable for transmission by all means of communication coming into operation 1 April 1969.

This edition incorporates Amendment List No 1 (effective from 1 January 1972) Amendment List No 2 (effective from 1 January 1974) Amendment List No 3 (effective from 1 January 1980) Amendment List No 4 (effective from 1 January 1986) Amendment List No 5 (effective from 1 January 1988) and Amendment List No 6.

The Code was adopted by the fourth Assembly of IMCO in 1965.

Contents

Distress signals *page ix*
Alphabetical flags and numeral pendants *x*
Table of life-saving signals *xii*
Radiotelephone procedures *xiv*

Chapter I	**Explanation and general remarks**	*1*
Chapter II	**Definitions**	*2*
Chapter III	**Methods of signalling**	*3*
Chapter IV	**General instructions**	*4*
Chapter V	**Flag signalling**	*7*
Chapter VI	**Flashing light signalling**	*9*
Chapter VII	**Radiotelephony**	*11*
Chapter VIII	**Morse signalling by hand-flags or arms**	*13*
Chapter IX	**Sound signalling**	*16*
Chapter X	**Morse symbols—phonetic tables—procedure signals**	*17*
Chapter XI	**Single-letter signals**	*21*
Chapter XII	**Single-letter signals with complements**	*23*
Chapter XIII	**Single-letter signals between ice-breaker and assisted vessels**	*24*
Chapter XIV	**Identification of medical transport in armed conflicts**	*26*

General section

Part I **Distress—emergency**
Abandon *29*
Accident—doctor—injured/sick *29*
Aircraft—helicopter *31*
Assistance *34*
Boats—rafts *36*
Disabled—drifting—sinking *38*
Distress *39*
Position *41*
Search and rescue *43*
Survivors *47*

Part II **Casualties—damages**
Collision *48*
Damages—repairs *48*
Diver—under-water operations *49*
Fire—explosion *50*
Grounding—beaching—refloating *51*
Leak *53*
Towing—tugs *54*

Part III **Aids to navigation—navigation—hydrography**
Aids to navigation *59*
Bar *59*
Bearings *59*
Canal—channel—fairway *60*
Course *61*
Dangers to navigation—warnings *62*
Depth—draught *65*
Electronic navigation *67*

Part III **Aids to navigation—navigation—hydrography** *continued*
Mines—minesweeping *page 68*
Navigation lights—searchlight *69*
Navigating and steering instructions *69*
Tide *71*

Part IV **Manoeuvres**
Ahead—astern *72*
Alongside *72*
To anchor—anchor(s)—anchorage *73*
Engines—propeller *74*
Landing—boarding *75*
Manoeuvre *75*
Proceed—under way *75*
Speed *77*
Stop—heave to *78*

Part V **Miscellaneous**
Cargo—ballast *79*
Crew—persons on board *79*
Fishery *80*
Pilot *81*
Port—harbour *82*
Miscellaneous *82*

Part VI **Meteorology—weather**
Clouds *84*
Gale—storm—tropical storm *84*
Ice—icebergs *84*
Ice-breaker *86*
Atmospheric pressure—temperature *87*
Sea—swell *87*
Visibility—fog *89*
Weather—weather forecast *89*
Wind *90*

Part VII **Routing of Ships**

Part VIII **Communications**
Acknowledge—answer *91*
Calling *91*
Cancel *91*
Communicate *91*
Exercise *93*
Reception—transmission *93*
Repeat *93*

Part IX **International sanitary regulations**
Pratique messages *94*

Tables of complements *95*

Medical section

Instructions *page 99*

Part I Request for medical assistance
Request—general information *102*
Description of patient *102*
Previous health *103*
Localization of symptoms, diseases or injuries *103*
General symptoms *103*
Particular symptoms *107*
Accidents, injuries, fractures, suicides and poisons *107*
Diseases of nose and throat *108*
Diseases of respiratory system *109*
Diseases of digestive system *110*
Diseases of the genito-urinary system *111*
Diseases of the nervous system and mental diseases *112*
Diseases of the heart and circulatory system *113*
Infectious and parasitic diseases *113*
Venereal diseases *114*
Diseases of the ear *114*
Diseases of the eye *115*
Diseases of the skin *115*
Diseases of muscles and joints *115*
Miscellaneous illnesses *116*
Childbirth *116*
Progress report *117*

Part II Medical advice
Request for additional information *118*
Diagnosis *118*
Special treatment *118*
Treatment by medicaments *120*
Prescribing *120*
Method of administration and dose *120*
Frequency of dose *120*
Frequency of external application *121*
Diet *121*
Childbirth *121*
General instructions *122*

Tables of complements
Table M I Regions of the body *125*
Table M II List of common diseases *128*
Table M III List of medicaments *129*

Medical index *page 133*

General index *139*

Illustrations

Alphabetical flags and numeral pendants *x*
Life-saving signals *xii*
Morse signalling by hand-flags or arms *14*
Regions of the body—front *124*
Regions of the body—back *126*

Distress signals

*Prescribed by the International Regulations for
Preventing Collisions at Sea, 1972*

1. The following signals, used or exhibited either together or separately, indicate distress and need of assistance:
   (a) a gun or other explosive signal fired at intervals of about a minute;
   (b) a continuous sounding with any fog-signalling apparatus;
   (c) rockets or shells, throwing red stars fired one at a time at short intervals;
   (d) a signal made by radiotelegraphy or by any other signalling method consisting of the group ... — — — ... (SOS) in the Morse Code;
   (e) a signal sent by radiotelephony consisting of the spoken word "Mayday";
   (f) the International Code Signal of distress indicated by N.C.;
   (g) a signal consisting of a square flag having above or below it a ball or anything resembling a ball;
   (h) flames on the vessel (as from a burning tar barrel, oil barrel, etc.);
   (i) a rocket parachute flare or a hand flare showing a red light;
   (j) a smoke signal giving off orange-coloured smoke;
   (k) slowly and repeatedly raising and lowering arms outstretched to each side;
   (l) the radiotelegraph alarm signal*;
   (m) the radiotelephone alarm signal**;
   (n) signals transmitted by emergency position-indicating radio beacons***.

2. The use or exhibition of any of the foregoing signals except for the purpose of indicating distress and need of assistance and the use of other signals which may be confused with any of the above signals is prohibited.

3. Attention is drawn to the relevant section of the Merchant Ship Search and Rescue Manual and the following signals:
   (a) a piece of orange-coloured canvas with either a black square and circle or other appropriate symbol (for identification from the air);
   (b) a dye-marker.'

---

* A series of twelve four-second dashes with intervals of one second.
** Two audio tones transmitted alternately at a frequency of 2200 Hz and 1300 Hz for a duration of 30 seconds to one minute.
*** Either the signal described in ** above or a series of single tones at a frequency of 1300 Hz.

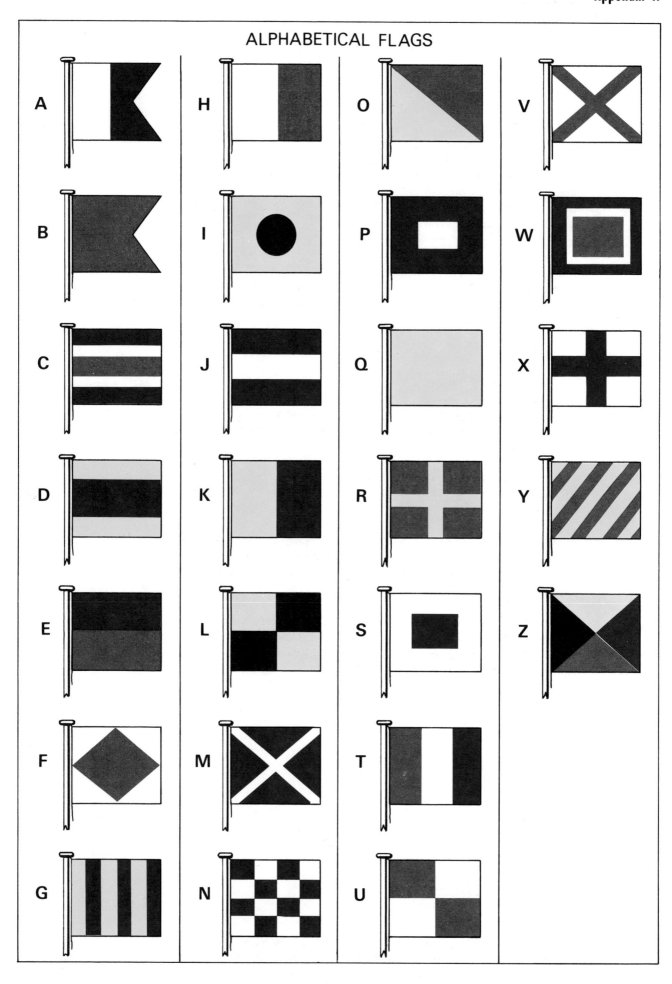

Appendix A (continued)

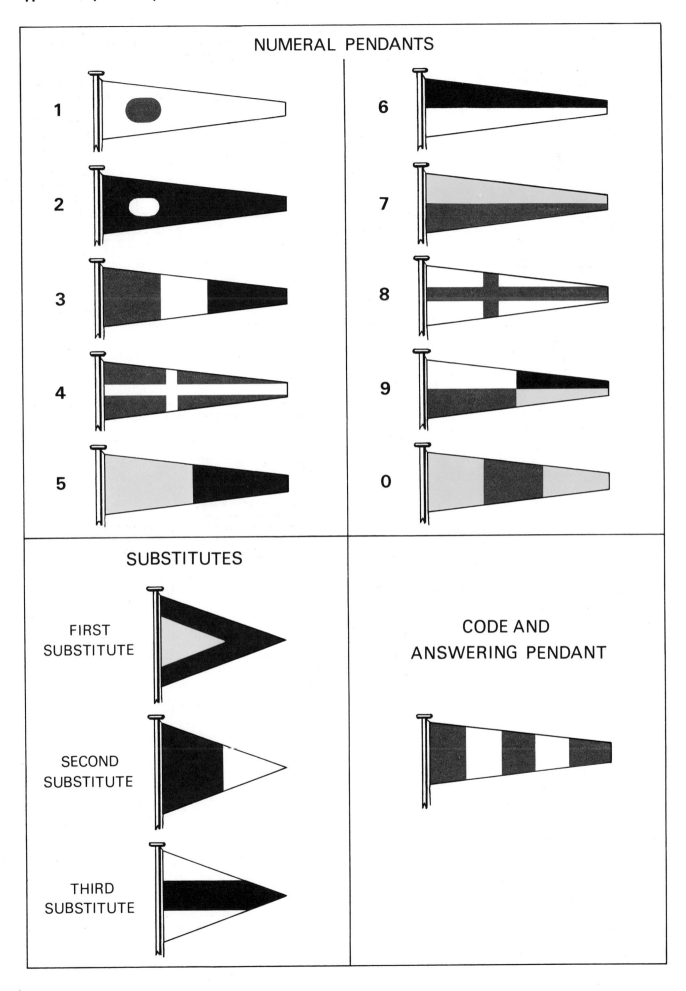

Life-Saving Signals

Appendix B

Signals to be employed in connexion with the use of shore life-saving apparatus

	MANUAL SIGNALS	LIGHT SIGNALS	OTHER SIGNALS	SIGNIFICATION
Day signals	Vertical motion of a white flag or of the arms	or firing of a green star signal		In general: affirmative Specifically: rocket line is held – tail block is made fast – hawser is made fast – man is in the breeches buoy – haul away
Night signals	Vertical motion of a white light or flare	or firing of a green star signal		
Day signals	Horizontal motion of a white flag or of the arms extended horizontally	or firing of a red star signal		In general: negative Specifically: slack away – avast hauling
Night signals	Horizontal motion of a white light or flare	or firing of a red star signal		

Replies from life-saving stations or maritime rescue units to distress signals made by a ship or person

	MANUAL SIGNALS	LIGHT SIGNALS	OTHER SIGNALS	SIGNIFICATION
Day signals		Orange smoke signal	or combined *light and sound* signal (thunder-light) consisting of 3 single signals which are fed at intervals of approximately one minute	**You are seen – assistance will be given as soon as possible** (Repetition of such signal shall have the same meaning)
Night signals		White star rocket consisting of 3 single signals which are fired at intervals of approximately one minute		

If necessary, the day signals may be given at night or the night signals by day

Appendix C — Life-Saving Signals

Landing signals for the guidance of small boats with crews or persons in distress

	MANUAL SIGNALS	LIGHT SIGNALS	OTHER SIGNALS	SIGNIFICATION
Day signals	**Vertical** motion of a white flag or of the arms	or firing of a green star signal	— — • — — or code letter **K** given by light or sound-signal apparatus	This is the best place to land
Night signals	**Vertical** motion of a white light or flare	or firing of a green star signal	— — • — — or code letter **K** given by light or sound-signal apparatus	

A range (indication of direction) may be given by placing a steady white light or flare at a lower level and in line with the observer

	MANUAL SIGNALS	LIGHT SIGNALS	OTHER SIGNALS	SIGNIFICATION
Day signals	**Horizontal** motion of a white flag or of the arms extended horizontally	or firing of a red star signal	• • • or code letter **S** given by light or sound-signal apparatus	Landing here highly dangerous
Night signals	**Horizontal** motion of a light or flare	or firing of a red star signal	• • • or code letter **S** given by light or sound-signal apparatus	
Day signals	1 **Horizontal** motion of a white flag followed by 2 the placing of the white flag in the ground and 3 by the carrying of another white flag in the direction to be indicated	1 or firing of a red star signal vertically and 2 a white star signal in the direction towards the better landing place	1 or signalling the code letter **S** (...) followed by the code letter **R** (.—.) if a better landing place for the craft in distress is located more to the *right* in the direction of approach 2 or signalling the code letter **S** (...) followed by the code letter **L** (.—..) if a better landing place for the craft in distress is located more to the *left* in the direction of approach	Landing here highly dangerous. A more favourable location for landing is in the direction indicated
Night signals	1 **Horizontal** motion of a white light or flare 2 followed by the placing of the white light or flare on the ground and 3 the carrying of another white light or flare in the direction to be indicated	1 or firing of a red star signal vertically and 2 white star signal in the direction towards the better landing place	1 or signalling the code letter **S** (...) followed by the code letter **R** (.—.) if a better landing place for the craft in distress is located more to the *right* in the direction of approach 2 or signalling the code letter **S** (...) followed by the code letter **L** (.—..) if a better landing place for the craft in distress is located more to the *left* in the direction of approach	

Appendix D

Air-surface visual signals.

Signals used by aircraft engaged in search and rescue operations to direct ships towards an aircraft ship or person in distress.

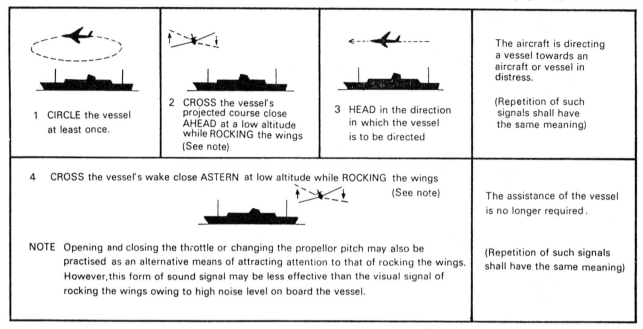

PROCEDURES PERFORMED IN SEQUENCE BY AN AIRCRAFT	SIGNIFICATION
1. CIRCLE the vessel at least once. — 2. CROSS the vessel's projected course close AHEAD at a low altitude while ROCKING the wings (See note) — 3. HEAD in the direction in which the vessel is to be directed	The aircraft is directing a vessel towards an aircraft or vessel in distress. (Repetition of such signals shall have the same meaning)
4. CROSS the vessel's wake close ASTERN at low altitude while ROCKING the wings (See note). NOTE: Opening and closing the throttle or changing the propellor pitch may also be practised as an alternative means of attracting attention to that of rocking the wings. However, this form of sound signal may be less effective than the visual signal of rocking the wings owing to high noise level on board the vessel.	The assistance of the vessel is no longer required. (Repetition of such signals shall have the same meaning)

Signals used by a vessel in response to an aircraft engaged on search and rescue operations.

			SIGNIFICATION
Hoist "Code and Answering" pendant Close up; or	Change the heading to the required direction; or	Flash Morse Code signal "T" by signal lamp.	Acknowledges receipt of aircraft's signal
Hoist international flag "N" (NOVEMBER). or		Flash Morse Code signal "N" by signal lamp.	Indicates inability to comply

Signals to survivors

Procedures performed by an aircraft   SIGNIFICATION.

Drop a message or	Drop communication equipment suitable for establishing direct contact		The aircraft wishes to inform or instruct survivors

Signals used by survivors in response to
a message dropped by an aircraft   SIGNIFICATION

▬ or • ▬ • Flash Morse code signal 'T' or 'R' by light or	Use any other suitable signal		Dropped message is understood by the survivors
•▬• •▬ ▬• ▬ Flash Morse code signal 'RPT' by light			Dropped message is not understood by the survivors

*   High visibility coloured streamer

Surface to air visual signals

Communication from surface craft or survivors to an aircraft.

Use the following surface-to-air visual signals by displaying the appropriate signal on the deck or on the ground	

Message	ICAO*/IMO** visual signals
– Require assistance	V
– Require medical assistance	X
– No or negative	N
– Yes or affirmative	Y
– Proceeding in this direction	↑

* ICAO annex 12 – Search and rescue
** IMOSAR and MERSAR Manuals"

Reply from an aircraft observing the above signals from a surface craft or survivors SIGNIFICATION

Drop a message or	Rock the Wings (during the daylight) or	Flash the landing lights or navigation lights on and off twice (during hours of darkness) or	Flash Morse code signal 'T' or 'R' by light or	Use any other suitable signal	Message understood
Fly straight and level without rocking wings or	Flash Morse code signal 'RPT' by light or	Use any other suitable signal			Message not understood (repeat)

Radiotelephone procedures

Reception of safety messages

Any message which you hear prefixed by one of the following words concerns SAFETY

MAYDAY (Distress) — Indicates that a ship, aircraft or other vehicle is threatened by grave and imminent danger and requests immediate assistance.

PAN (Urgency) — Indicates that the calling station has a very urgent message to transmit concerning the safety of a ship, aircraft or other vehicle, or the safety of a person.

SÉCURITÉ (Safety) — Indicates that the station is about to transmit a message concerning the safety of navigation or giving important meteorological warnings.

If you hear these words, pay particular attention to the message and call the master or the officer on watch.

Table 1

Phonetic Alphabet and Figure Spelling Tables

(May be used when transmitting plain language or code)

Letter	Word	Pronounced as	Letter	Word	Pronounced as
A	Alfa	**AL** FAH	N	November	NO **VEM** BER
B	Bravo	**BRAH** VOH	O	Oscar	**OSS** CAH
C	Charlie	**CHAR** LEE or **SHAR** LEE	P	Papa	PAH **PAH**
			Q	Quebec	KEH **BECK**
D	Delta	**DELL** TAH	R	Romeo	**ROW** ME OH
E	Echo	**ECK** OH	S	Sierra	SEE **AIR** RAH
F	Foxtrot	**FOKS** TROT	T	Tango	**TANG** GO
G	Golf	GOLF	U	Uniform	**YOU** NEE FORM or **OO** NEE FORM
H	Hotel	HOH **TELL**	V	Victor	**VIK** TAH
I	India	**IN** DEE AH	W	Whiskey	**WISS** KEY
J	Juliett	**JEW** LEE **ETT**	X	X-ray	**ECKS** RAY
K	Kilo	**KEY** LOH	Y	Yankee	**YANG** KEY
L	Lima	**LEE** MAH	Z	Zulu	**ZOO** LOO
M	Mike	MIKE			

Note: The syllables to be emphasized are underlined.

Figure or mark to be transmitted	Word	Pronounced as	Figure or mark to be transmitted	Word	Pronounced as
0	NADAZERO	NAH-DAH-ZAY-ROH	6	SOXISIX	SOK-SEE-SIX
1	UNAONE	OO-NAH-WUN	7	SETTESEVEN	SAY-TAY-SEVEN
2	BISSOTWO	BEES-SOH-TOO	8	OKTOEIGHT	OK-TOH-AIT
3	TERRATHREE	TAY-RAH-TREE	9	NOVENINE	NO-VAY-NINER
4	KARTEFOUR	KAR-TAY-FOWER	Decimal point	DECIMAL	DAY-SEE-MAL
5	PANTAFIVE	PAN-TAH-FIVE	Full stop	STOP	STOP

Note: Each syllable should be equally emphasized.

Examples of distress procedure

1. Where possible, transmit **Alarm Signal** followed by spoken words Mayday Mayday Mayday. This is . . . (name of ship spoken three times, or call sign of ship spelt three times using Table 1) Mayday . . . (name or call sign of ship) Position 54 25 North 016 33 West I am on fire and require immediate assistance.

2. Where possible, transmit **Alarm Signal** followed by spoken words Mayday Mayday Mayday . . . (name of ship spoken three times, or call sign of ship spelt three times using Table 1) Mayday . . . (name or call sign of ship) Interco Alfa Nadazero Unaone Pantafive Ushant Romeo Kartefour Nadazero Delta X-ray. "(Ship) in Distress Position 015 Degrees Ushant 40 miles I am sinking."

3. Where possible, transmit **Alarm Signal** followed by spoken words Mayday Mayday Mayday . . . (name of ship spoken three times, or call sign of ship spelt three times using Table 1) Mayday . . . (name or call sign of ship) Interco Lima Pantafive Kartefour Bissotwo Pantafive November Golf Nadazero Unaone Soxisix Terrathree Terrathree Whiskey Charlie Bravo Soxisix. "(Ship) in Distress Position Latitude 54 25 North Longitude 016 33 West I require immediate assistance (I am on fire)."

Radiotelephone procedures

Name of ship . . . .
Call sign . . . .

Distress transmitting procedures

To be used only if **Immediate Assistance** is required

Use plain language whenever possible. If language difficulties are likely to arise use Tables 2 and 3 below, sending the word INTERCO to indicate that the message will be in the International Code of Signals.

Call out letters in Table 1. Call out numbers figure by figure as in Table 1.

To indicate Distress.

1. If possible transmit the Alarm Signal (i.e. two-tone signal) for 30 seconds to one minute, but do not delay the message if there is insufficient time in which to transmit the Alarm Signal.

2. Send the following Distress Call:
   Mayday Mayday Mayday.. This is . . . (name or call sign of ship spoken three times).

3. Then send the Distress Message composed of:
   Mayday followed by the name or call sign of ship;
   Position of ship;
   Nature of distress;
   And, if necessary, transmit the nature of the aid required and any other information which will help the rescue.

Table 2

Position in Code from the International Code of Signals

(1) **By Bearing and Distance from a Landmark**
Code letter A (Alfa) followed by a three-figure group for ship's TRUE bearing from landmark;

Name of landmark;
Code letter R (Romeo) followed by one or more figures for distance in nautical miles.

or

(2) **By Latitude and Longitude**

Latitude
Code letter L (Lima) followed by a four-figure group; (2 figures for Degrees, 2 figures for Minutes) and either—N (November) for Latitude North, or S (Sierra) for Latitude South.

Longitude
Code letter G (Golf) followed by a five-figure group; (3 figures for Degrees, 2 figures for Minutes) and either—E (Echo) for Longitude East, or W (Whiskey) for Longitude West.

Table 3

Nature of Distress in Code from the International Code of Signals

Code Letters	Words to be transmitted	Text of Signal
AE	Alfa Echo	I must abandon my vessel.
BF	Bravo Foxtrot	Aircraft is ditched in position indicated and requires immediate assistance.
CB	Charlie Bravo	I require immediate assistance.
CB6	Charlie Bravo Soxisix	I require immediate assistance, I am on fire.
DX	Delta X-ray	I am sinking.
HW	Hotel Whiskey	I have collided with surface craft.

Answer to Ship in Distress

CP	Charlie Papa	I am proceeding to your assistance.
ED	Echo Delta	Your distress signals are understood.
EL	Echo Lima	Repeat the distress position.

Note: A more comprehensive list of signals may be found in the International Code of Signals.

Chapter I Explanation and general remarks

1. The purpose of the International Code of Signals is to provide ways and means of communication in situations related essentially to safety of navigation and persons, especially when language difficulties arise. In the preparation of the Code, account was taken of the fact that wide application of radiotelephony and radiotelegraphy can provide simple and effective means of communication in plain language whenever language difficulties do not exist.

2. The signals used consist of:
(a) Single-letter signals allocated to significations which are very urgent, important, or of very common use;
(b) Two-letter signals for the General Section;
(c) Three-letter signals beginning with "M" for the Medical Section.

3. The Code follows the basic principle that each signal should have a complete meaning. This principle is followed throughout the Code; in certain cases complements are used, where necessary, to supplement the available groups.

4. Complements express:
(a) Variations in the meaning of the basic signal.
Examples:
"CP" = "I am (or vessel indicated is) proceeding to your assistance."
"CP 1" = "SAR aircraft is coming to your assistance."
(b) Questions concerning the same basic subject or basic signal.
Examples:
"DY" = "Vessel (name or identity signal) has sunk in lat . . . long . . .".
"DY 4" = "What is the depth of water where vessel sunk?"
(c) Answers to a question or request made by the basic signal.
Examples:
"HX" = "Have you received any damage in collision?"
"HX 1" = "I have received serious damage above the water-line".
(d) Supplementary, specific or detailed information.
Examples:
"IN" = "I require a diver".
"IN 1" = "I require a diver to clear propeller".

5. Complements appearing in the text more than once have been grouped in three tables. These tables should be used only as and when specified in the text of the signals.

6. Text in brackets indicates:
(a) an alternative, e.g.: ". . . (or survival craft) . . .";
(b) information which may be transmitted if it is required or if it is available, e.g.: ". . . (position to be indicated if necessary)";
(c) an explanation of the text.

7. The material is classified according to subject and meaning. Extensive cross referencing of the signals in the right-hand column is used to facilitate coding.

Chapter II Definitions

For the purpose of this Code the following terms shall have the meanings defined below:

Addressee is the authority to whom a signal is addressed.

Group denotes one or more continuous letters and/or numerals which together compose a signal.

A **hoist** consists of one or more groups displayed from a single halyard. A hoist or signal is said to be **at the dip** when it is hoisted about half of the full extent of the halyards. A hoist or signal is said to be **close up** when it is hoisted to the full extent of the halyards.

Identity signal or call sign is the group of letters and figures assigned to each station by its administration.

A **numeral group** consists of one or more numerals.

Originator is the authority who orders a signal to be sent.

Procedure denotes the rules drawn up for the conduct of signalling.

Procedure signal is a signal designed to facilitate the conduct of signalling (see pages 19–20).

Receiving station is the station by which a signal is actually being read.

Sound signalling is any method of passing Morse signals by means of siren, whistle, foghorn, bell, or other sound apparatus.

Station means a ship, aircraft, survival craft or any place at which communications can be effected by any means.

Station of destination is that station in which the signal is finally received by the addressee.

Station of origin is that station where the originator hands in a signal for transmission, irrespective of the method of communication employed.

Tackline is a length of halyard about six feet (2 m) long, used to separate each group of flags.

Time of origin is the time at which a signal is ordered to be made.

Transmitting station is the station by which a signal is actually being made.

Visual signalling is any method of communication, the transmission of which is capable of being seen.

Chapter III **Methods of signalling**

1. The methods of signalling which may be used are:
(a) Flag signalling, the flags used being those shown on page x.
(b) Flashing light signalling, using the Morse symbols shown on page 17.
(c) Sound signalling, using the Morse symbols shown on page 17.
(d) Voice over a loud hailer.
(e) Radiotelegraphy.
(f) Radiotelephony.
(g) Morse signalling by hand flags or arms.

Flag signalling

2. A set of signal flags consists of twenty-six alphabetical flags, ten numeral pendants, three substitutes and the answering pendant. Detailed instructions for signalling by flags are given in chapter V.

Flashing light and sound signalling

3. The morse symbols representing letters, numerals, etc, are expressed by dots and dashes which are signalled either singly or in combination. The dots and dashes and spaces between them should be made to bear the following ratio, one to another, as regards their duration:
(a) A dot is taken as the unit;
(b) A dash is equivalent to three units;
(c) The space of time between any two elements of a symbol is equivalent to one unit; between two complete symbols it is equivalent to three units and between two words or groups it is equivalent to seven units.

4. In flashing light and sound signalling, while generally obeying the instructions laid down here, it is best to err on the side of making the dots rather shorter in their proportion to the dashes as it then makes the distinction between the elements plainer. The standard rate of signalling by flashing light is to be regarded as forty letters per minute. Detailed instructions for signalling by flashing light and sound are given in chapters VI and IX

Voice over a loud hailer

5. Whenever possible plain language should be used but where a language difficulty exists groups from the International Code of Signals could be transmitted using the phonetic spelling tables.

Radiotelegraphy and Radiotelephony

6. When radiotelegraphy or radiotelephony is used for the transmission of signals, operators should comply with the Radio Regulations of the International Telecommunication Union then in force (see chapter VII on **Radiotelephony**).

Chapter IV General instructions

Originator and addressee of message 1. Unless otherwise indicated all signals between vessels are made from the master of the vessel of origin to the master of the vessel of destination.

Identification of ships and aircraft 2. Identity signals for ships and aircraft are allocated on an international basis. The identity signal may therefore indicate the nationality of a ship or aircraft.

Use of identity signals 3. Identity signals may be used for two purposes:
(a) to speak to, or call, a station;
(b) to speak of, or indicate, a station.
Examples:
"YP LABC" = "I wish to communicate with vessel LABC by . . . (Complements table I)".
"HY 1 LABC" = "The vessel LABC with which I have been in collision has resumed her voyage".

Names of vessels and/or places 4. Names of vessels and/or places are to be spelt out.
Example:
"RV Gibraltar" = "You should proceed to Gibraltar".

How to signal numbers 5. (a) Numbers are to be signalled as follows:
(i) flag signalling: by the numeral pendants of the Code.
(ii) flashing light or sound signalling: usually by the numerals in the morse code; they may also be spelt out.
(iii) radiotelephony or loud hailer: by the code words of the figure-spelling table.
(b) Figures which form part of the basic signification of a signal are to be sent together with the basic group.
Examples:
"DI20" = "I require boats for 20 persons".
"FJ2" = "Position of accident (or survival craft) is marked by sea marker".
(c) A decimal point between numerals is to be signalled as follows:
(i) flag signalling: by inserting the answering pendant where it is desired to express the decimal point.
(ii) flashing light and sound signalling: by "decimal point" signal "$\overline{AAA}$".
(iii) voice: by use of the word "DECIMAL" as indicated in the figure-spelling table.
(d) Wherever the text allows depths, etc., to be signalled in feet or in metres, the figures should be followed by "F" to indicate feet or by "M" to indicate metres.

General instructions   5

Azimuth or bearing
6. They are to be expressed in three figures denoting degrees from 000 to 359, measured clockwise. If there is any possibility of confusion, they should be preceded by the letter "A". They are always to be true unless expressly stated to be otherwise in the context.
Examples:
"LW 005" = "I receive your transmission on bearing 005°".
"LT A120 T1540" = "Your bearing from me is 120° at (local time) 1540"

Course
7. Course is to be expressed in three figures denoting degrees from 000 to 359, measured clockwise. If there is any possibility of confusion, they should be preceded by the letter "C". They are always to be true unless expressly stated to be otherwise in the context.
Examples:
"MD 025" = "My course is 025°".
"GR C240 S18" = "Vessel coming to your rescue is steering course 240°, speed 18 knots".

Date
8. Dates are to be signalled by two, four or six figures preceded by the letter "D". The first two figures indicate the day of the month. When they are used alone they refer to the current month. Example: "D15" transmitted on the 15th or any other date in April means "15 April". The two figures which follow indicate the month of the year. Example: "D1504" means "15 April". Where necessary the year may be indicated by two further figures. Example: "D181063" means "18 October 1963".

Latitude
9. Latitude is expressed by four figures preceded by the letter "L". The first two figures denote the degrees and the last two the minutes. The letters "N" (North) or "S" (South) follow if they are needed; however, for reasons of simplicity they may be omitted if there is no risk of confusion.
Example:
"L3740S" = "Latitude 37° 40'S".

Longitude
10. Longitude is expressed by four or, if necessary, five figures preceded by the letter "G". The first two (or three) figures denote the degrees and the last two the minutes. When the longitude is more than 99°, no confusion will normally arise if the figure indicating hundreds of degrees is omitted. However, where it is necessary to avoid confusion the five figures should be used. The letters "E" (East) or "W" (West) follow if they are needed, otherwise they may be omitted, as in the case of latitude.
Example:
"G13925E" = "Longitude 139° 25'E".
A signal requiring the indication of position to complete its signification should be signalled as follows:
"CH L2537N G4015W" = "Vessel indicated is reported as requiring assistance in lat 25° 37'N long 40° 15'W".

Distance 11. Figures preceded by the letter "R" indicate distance in nautical miles.
Example:
"OVA080 R10" = "Mine(s) is(are) believed to be bearing 080° from me, distance 10 miles".
The letter "R" may be omitted if there is no possibility of confusion.

Speed 12. Speed is indicated by figures preceded by:
(a) the letter "S" to denote speed in knots, or
(b) the letter "V" to denote speed in kilometres per hour.
Examples:
"BQ S300" = "The speed of my aircraft in relation to the surface of the earth is 300 knots".
"BQ V300" = "The speed of my aircraft in relation to the surface of the earth is 300 kilometres per hour".

Time 13. Times are to be expressed in four figures, of which the first two denote the hour (from 00 = midnight up to 23 = 11 p.m.), and the last two denote the minutes (from 00 to 59). The figures are preceded by:
(a) the letter "T" indicating "Local time", or
(b) the letter "Z" indicating "Greenwich Mean Time".
Examples:
"BH T1045 L2015N G3840W C125 = "I sighted an aircraft at local time 1045 in lat 20° 15'N long 38°40'W flying on course 125°".
"RX Z0830" = "You should proceed at GMT 0830".

Time of origin 14. The time of origin may be added at the end of the text. It should be given to the nearest minute and expressed by four figures. Apart from indicating at what time a signal originated it also serves as a convenient reference number.

Communication by local signal codes 15. If a vessel or a coast station wishes to make a signal in a local code, the signal "YV 1" = "The groups which follow are from the local code" should precede the local signal, if it is necessary, in order to avoid misunderstanding.

Chapter V Flag signalling

1. As a general rule only one hoist should be shown at a time. Each hoist or group of hoists should be kept flying until it has been answered by the receiving station (see paragraph 3). When more groups than one are shown on the same halyard they must be separated by a tackline. The transmitting station should always hoist the signal where it can be most easily seen by the receiving station, that is, in such a position that the flags will blow out clear and be free from smoke.

How to call

2. The identity signal of the station(s) addressed is to be hoisted with the signal (see chapter IV, paragraph 3). If no identity signal is hoisted it will be understood that the signal is addressed to all stations within visual signalling distance. If it is not possible to determine the identity signal of the station to which it is desired to signal, the group
"VF" ="You should hoist your identity signal" or
"CS" ="What is the name or identity signal of your vessel (or station)?" should be hoisted first; at the same time the station will hoist its own identity signal. The group "YQ" ="I wish to communicate by . . . (Complements table I) with vessel bearing . . . from me" can also be used.

How to answer signals

3. All stations to which signals are addressed or which are indicated in signals are to hoist the answering pendant at the dip as soon as they see each hoist and close up immediately they understand it; it is to be lowered to the dip as soon as the hoist is hauled down in the transmitting station, being hoisted close up again as soon as the next hoist is understood.

How to complete a signal

4. The transmitting station is to hoist the answering pendant singly after the last hoist of the signal to indicate that the signal is completed. The receiving station is to answer this in a similar manner to all other hoists (see paragraph 3).

How to act when signals are not understood

5. If the receiving station cannot clearly distinguish the signal made to it, it is to keep the answering pendant at the dip. If it can distinguish the signal but cannot understand the purport of it, it can hoist the following signals:
"ZQ" ="Your signal appears incorrectly coded. You should check and repeat the whole", or "ZL" ="Your signal has been received but not understood".

The use of substitutes

6. The use of substitutes is to enable the same signal flag— either alphabetical flag or numeral pendant—to be repeated one or more times in the same group, in case only one set of flags is carried on board. The first substitute always repeats the uppermost signal flag of that class of flags which immediately precedes the substitute. The second substitute always repeats the second and the third substitute repeats the third signal flag, counting from the top of that class of flags which immediately precedes them. No substitute can ever be used more than once

in the same group. The answering pendant when used as a decimal point is to be disregarded in determining which substitute to use. Example:

The signal "VV" would be made as follows:
>V
>first substitute.

The number "1100" would be made by numeral pendants as follows:
>1
>first substitute
>0
>third substitute.

The signal "L2330" would be made as follows:
>L
>2
>3
>second substitute
>0

In this case, the second substitute follows a numeral pendant and therefore it can only repeat the second numeral in the group.

How to spell 7. Names in the text of a signal are to be spelt out by means of the alphabetical flags. The signal "YZ" ="The words which follow are in plain language" can be used, if necessary.

'Use of the Code Pendant by ships of war 8. When a ship of war wishes to communicate with a merchant vessel she will hoist the Code Pendant in a conspicuous position, and keep it flying during the whole of the time the signal is being made.

Chapter VI **Flashing light signalling**

1. A signal made by flashing light is divided into the following parts:
   (a) The call. It consists of the general call or the identity signal of the station to be called. It is answered by the answering signal.
   (b) The identity. The transmitting station makes "DE" followed by its identity signal or name. This will be repeated back by the receiving station which then signals its own identity signal or name. This will also be repeated back by the transmitting station.
   (c) The text. This consists of plain language or code groups. When code groups are to be used they should be preceded by the signal "YU". Words of plain language may also be in the text, when the signal includes names, places, etc. Receipt of each word or group is acknowledged by "T".
   (d) The ending. It consists of the ending signal "$\overline{AR}$" which is answered by "R".

2. If the entire text is in plain language the same procedure is to be followed. The call and identity may be omitted when two stations have established communications and have already exchanged signals.

3. A list of procedure signals appears in pages 19 and 20. Although the use of these signals is self-explanatory, the following notes might be found useful:
   (a) **The General call signal** (or call for unknown station) "$\overline{AA}$ $\overline{AA}$ $\overline{AA}$" etc., is made to attract attention when wishing to signal to all stations within visual signalling distance or to a station whose name or identity signal is not known. The call is continued until the station addressed answers.
   (b) The **Answering signal** "$\overline{TTTT}$" etc., is made to answer the call and it is to be continued until the transmitting station ceases to make the call. The transmission starts with the signal "DE" followed by the name or identity signal of the transmitting station.
   (c) The letter "T" is used to indicate the receipt of each word or group.
   (d) The **Erase signal** "$\overline{EEEEEE}$" etc, is used to indicate that the last group or word was signalled incorrectly. It is to be answered with the erase signal. When answered, the transmitting station will repeat the last word or group which was correctly signalled and then proceed with the remainder of the transmission.
   (e) The **Repeat signal** "RPT" is to be used as follows:
      (i) by the transmitting station to indicate that it is going to repeat ("I repeat"). If such a repetition does not follow immediately after "RPT", the signal should be interpreted as a request to the receiving station to repeat the signal received ("Repeat what you have received").
      (ii) by the receiving station to request for a repetition of the signal transmitted ("Repeat what you have sent").
      (iii) The special **Repetition signals** "AA", "AB", "WA", "WB" and "BN" are made by the receiving station as appropriate. In each case they are made immediately after the repeat signal "RPT".

Examples:
"RPT AB KL" = "Repeat all before group KL".
"RPT BN 'boats' 'survivors' " = "Repeat all between words 'boats' and 'survivors'".
If a signal is not understood, or, when decoded, it is not intelligible, the repeat signal is not used. The receiving station must then make the appropriate signal from the Code, e.g. "Your signal has been received but not understood".

(f) A correctly received **repetition is acknowledged** by the signal "OK". The same signal may be used as an affirmative answer to a question ("It is correct").

(g) **The Ending signal** "$\overline{AR}$" is used in all cases to indicate the end of a signal or the end of the transmission. The receiving station answers with the signal "R" = "Received" or "I have received your last signal".

(h) The transmitting station makes the signal "CS" when **requesting the name or identity signal** of the receiving station.

(i) The **Waiting signal or period signal** "$\overline{AS}$" is to be used as follows:
   (i) when made independently or after the end of a signal it indicates that the other station must wait for further communications (**waiting signal**);
   (ii) when it is inserted between groups it serves to separate them (**period signal**) to avoid confusion.

(j) The signal "C" should be used to indicate an affirmative statement or an affirmative reply to an interrogative signal; the signal "RQ" should be used to indicate a question. For a negative reply to an interrogative signal or for a negative statement, the signal "N" should be used in visual or sound signalling and the signal "NO" should be used for voice or radio transmission.
When the signals "N" or "NO" and "RQ" are used to change an affirmative signal into a negative statement or into a question, respectively, they should be transmitted after the main signal.
Examples:
"CY N" (or "NO" as appropriate) = "Boat(s) is(are) not coming to you."
"CW RQ" = "Is boat/raft on board?".
The signals "C", "N" or "NO" and "RQ" cannot be used in conjunction with single-letter signals.

Chapter VII **Radiotelephony**

1. When using the International Code of Signals in cases of language difficulties, the principles of the Radio Regulations of the International Telecommunication Union then in force have to be observed. Letters and figures are to be spelt in accordance with the spelling tables.

2. When coast and ship stations are called, the identity signals (call signs) or names shall be used.

Method of calling

3. The call consists of:
— the call sign or name of the station called, not more than three times at each call;
— the group "DE" (DELTA ECHO);
— the call sign or name of the calling station, not more than three times at each call.
Difficult names of stations should also be spelt. After contact has been established, the call sign or name need not be sent more than once.

Form of reply to calls

4. The reply to calls consists of:
— the call sign or name of the calling station, not more than three times;
— the group "DE" (DELTA ECHO);
— the call sign or name of the station called, not more than three times.

Calling all stations in the vicinity

5. The group "CQ" (CHARLIE QUEBEC) shall be used, but not more than three times at each call.

6. In order to indicate that Code groups of the International Code of Signals are following, the word "INTERCO" is to be inserted. Words of plain language may also be in the text when the signal includes names, places, etc. In this case the group "YZ" (YANKEE ZULU) is to be inserted if necessary.

7. If the station called is unable to accept traffic immediately, it should transmit the signal "$\overline{AS}$" (ALFA SIERRA), adding the duration of waiting time in minutes whenever possible.

8. The receipt of a transmission is indicated by the signal "R" (ROMEO).

9. If the transmission is to be repeated in total or in part, the signal "RPT" (ROMEO PAPA TANGO) shall be used, supplemented as necessary by:
AA (ALFA ALFA) = all after . . .
AB (ALFA BRAVO) = all before . . .
BN (BRAVO NOVEMBER) = all between . . . and . . .
WA (WHISKEY ALFA) = word or group after . . .
WB (WHISKEY BRAVO) = word or group before . . .

10. The end of a transmission is indicated by the signal "$\overline{AR}$" (ALFA ROMEO).

Chapter VIII Morse signalling by hand-flags or arms

Morse signalling by hand-flags or arms

1. Raising both hand-flags or arms

"dot"

2. Spreading out both hand-flags or arms at shoulder level

"dash"

3. Hand-flags or arms brought before the chest

Separation of "dots" and and/or "dashes"

4. Hand-flags or arms kept at 45° away from the body downwards

Separation of letters, groups or words

5. Circular motion of hand-flags or arms over the head
—erase signals, if made by the transmitting station.
—request for repetition if by the receiving station.

Note: The space of time between dots and dashes and between letters, groups or words should be such as to facilitate correct reception.

Morse Signalling 13
by hand-flags
or arms

1. A station which desires to communicate with another station by morse signalling by hand-flags or arms may indicate the requirement by transmitting to that station the signal "K2" by any method. The call signal "$\overline{AA}$ $\overline{AA}$ $\overline{AA}$" may be made instead.

2. On receipt of the call the station addressed should make the answering signal, or, if unable to communicate by this means, should reply with the signal "YS2" by any available method.

3. The call signal "$\overline{AA}$ $\overline{AA}$ $\overline{AA}$" and the signal "T" should be used respectively by the transmitting station and the addressed station.

4. Normally both arms should be used for this method of transmission but in cases where this is difficult or impossible one arm can be used.

5. All signals will end with the ending signal "$\overline{AR}$".

Chapter IX **Sound Signalling**

1. Owing to the nature of the apparatus used (whistle, siren, fog-horn, etc.) sound signalling is necessarily slow. Moreover, the misuse of sound signalling is of a nature to create serious confusion at sea. Sound signalling in fog should therefore be reduced to a minimum. Signals other than the single-letter signals should be used only in extreme emergency and never in frequented navigational waters.

2. The signals should be made slowly and clearly. They may be repeated, if necessary, but at sufficiently long intervals to ensure that no confusion can arise and that one-letter signals cannot be mistaken as two-letter groups.

3. Masters are reminded that the one-letter signals of the Code, which are marked*, when made by sound, may only be made in compliance with the requirements of the International Regulations for Preventing Collisions at Sea. Reference is also made to the single-letter signals provided for exclusive use between an ice-breaker and assisted vessels.

Chapter X Morse symbols – phonetic tables – procedure signals

Morse symbols

Alphabet

A .−	I ..	Q −−.−	Y −.−−
B −...	J .−−−	R .−.	Z −−..
C −.−.	K −.−	S ...	
D −..	L .−..	T −	
E .	M −−	U ..−	
F ..−.	N −.	V ...−	
G −−.	O −−−	W .−−	
H	P .−−.	X −..−	

Numerals

1 .−−−−	6 −....
2 ..−−−	7 −−...
3 ...−−	8 −−−..
4 ....−	9 −−−−.
5	0 −−−−−

Procedure signals

$\overline{\text{AR}}$.−.−.
$\overline{\text{AS}}$.−...
$\overline{\text{AAA}}$.−.−.−

Note: Certain letters, such as "ë", "ä", "ö" etc., have been omitted from this list of morse symbols because:
(a) they are not to be used internationally;
(b) they are contained in local codes, and
(c) some of them can be substituted by a combination of two letters.

16 Morse symbols
    phonetic tables
    procedure signals

Phonetic tables

for the pronunciation of letters and figures by radiotelephony or voice over a loud hailer

Letter-Spelling Table

letter	code word	pronounced as
A	Alfa	**AL** FAH
B	Bravo	**BRAH** VOH
C	Charlie	**CHAR** LEE (or **SHAR** LEE)
D	Delta	**DELL** TAH
E	Echo	**ECK** OH
F	Foxtrot	**FOKS** TROT
G	Golf	GOLF
H	Hotel	HOH **TELL**
I	India	**IN** DEE AH
J	Juliett	**JEW** LEE **ETT**
K	Kilo	**KEY** LOH
L	Lima	**LEE** MAH
M	Mike	MIKE
N	November	NO **VEM** BER
O	Oscar	**OSS** CAH
P	Papa	PAH **PAH**
Q	Quebec	KEH **BECK**
R	Romeo	**ROW** ME OH
S	Sierra	SEE **AIR** RAH
T	Tango	**TANG** GO
U	Uniform	**YOU** NEE FORM (or **OO** NEE FORM)
V	Victor	**VIK** TAH
W	Whiskey	**WISS** KEY
X	X-ray	**ECKS** RAY
Y	Yankee	**YANG** KEY
Z	Zulu	**ZOO** LOO

Note: The syllables to be emphasized are underlined

Figure-Spelling Table

figure or mark to be transmitted	code word	pronounced as
0	NADAZERO	NAH–DAH–ZAY–ROH
1	UNAONE	OO–NAH–WUN
2	BISSOTWO	BEES–SOH–TOO
3	TERRATHREE	TAY–RAH–TREE
4	KARTEFOUR	KAR–TAY–FOWER
5	PANTAFIVE	PAN–TAH–FIVE
6	SOXISIX	SOK–SEE–SIX
7	SETTESEVEN	SAY–TAY–SEVEN
8	OKTOEIGHT	OK–TOH–AIT
9	NOVENINE	NO–VAY–NINER
Decimal point	DECIMAL	DAY–SEE–MAL
Full stop	STOP	STOP

Note: Each syllable should be equally emphasized. The second component of each code word is the code word used in the Aeronautical Mobile Service.

Morse symbols 17
phonetic tables
procedure signals

Procedure signals

A bar over the letters composing a signal denotes that the letters are to be made as one symbol

1. Signals for voice transmissions (radiotelephony or loud hailer)

Signal	Pronounced as	Meaning
Interco	IN–TER–CO	International Code group(s) follow(s)
Stop	STOP	Full stop
Decimal	DAY–SEE–MAL	Decimal point
Correction	KOR REK SHUN	Cancel my last word or group. The correct word or group follows.

2. Signals for flashing-light transmission

$\overline{AA}\ \overline{AA}\ \overline{AA}$ etc.	Call for unknown station or general call
$\overline{EEEEE}$ etc.	Erase signal
$\overline{AAA}$	Full stop or decimal point
$\overline{TTTT}$ etc.	Answering signal
T	Word or group received.

3. Signals for flags, radiotelephony and radiotelegraphy transmissions

CQ   Call for unknown station(s) or general call to all stations

Note: When this signal is used in voice transmission, it should be pronounced in accordance with the letter-spelling table.

4. Signals for use where appropriate in all forms of transmission

AA "All after . . ." (used after the "Repeat signal" (RPT)), means "Repeat all after . . .".

AB "All before . . ." (used after the "Repeat signal" (RPT)) means "Repeat all before . . .".

$\overline{AR}$ Ending signal or End of Transmission or signal.

$\overline{AS}$ Waiting signal or period.

BN "All between . . . and . . ." (used after the "Repeat signal" (RPT)) means "Repeat all between . . . and . . .".

C Affirmative — YES or "The significance of the previous group should be read in the affirmative".

CS "What is the name or identity signal of your vessel (or station) ?".

DE "From . . ." (used to precede the name or identity signal of the calling station).

K "I wish to communicate with you" or "Invitation to transmit".

NO Negative — NO or "The significance of the previous group should be read in the negative". When used in voice transmission the pronunciation should be "NO".

OK Acknowledging a correct repetition or "It is correct".

RQ Interrogative, or, "The significance of the previous group should be read as a question".

R "Received" or "I have received your last signal".

RPT Repeat signal "I repeat" or "Repeat what you have sent" or "Repeat what you have received".

WA "Word or group after . . ." (used after the "Repeat signal" (RPT)) means "Repeat word or group after . . .".

WB "Word or group before . . ." (used after the "Repeat signal" (RPT)) means "Repeat word or group before . . .".

Notes:
(a) The procedure signals "C", "NO" and "RQ" cannot be used in conjunction with single-letter signals.
(b) Signals on COMMUNICATIONS appear on pages 91–93.
(c) When these signals are used by voice transmission the letters should be pronounced in accordance with the letter-spelling table, with the exception of "NO" which in voice transmission should be pronounced as "NO".

Chapter XI **Single-letter signals**

**May be made by any method of signalling.
For those marked* see note (1) below**

A   I have a diver down; keep well clear at slow speed.

*B   I am taking in, or discharging, or carrying dangerous goods.

*C   Yes (affirmative or "The significance of the previous group should be read in the affirmative").

*D   Keep clear of me; I am manoeuvring with difficulty.

*E   I am altering my course to starboard.

F   I am disabled; communicate with me.

*G   I require a pilot. When made by fishing vessels operating in close proximity on the fishing grounds it means: "I am hauling nets".

*H   I have a pilot on board.

*I   I am altering my course to port.

J   Keep well clear of me. I am on fire and have dangerous cargo on board, or I am leaking dangerous cargo.

K   I wish to communicate with you.

L   You should stop your vessel instantly.

M   My vessel is stopped and making no way through the water.

N   No (negative or "The significance of the previous group should be read in the negative"). This signal may be given only visually or by sound. For voice or radio transmission the signal should be "NO".

O   Man overboard.

P   **In harbour.** All persons should report on board as the vessel is about to proceed to sea.
At sea. It may also be used as a sound signal to mean: "I require a pilot".

Q   My vessel is 'healthy' and I request free pratique.

*S   I am operating astern propulsion.

*T   Keep clear of me; I am engaged in pair trawling.

U   You are running into danger.

V   I require assistance.

W   I require medical assistance.

X   Stop carrying out your intentions and watch for my signals.

Single-letter signals

Y   I am dragging my anchor.

*Z   I require a tug. When made by fishing vessels operating in close proximity on the fishing grounds it means: "I am shooting nets".

Notes:
1. Signals of letters marked* when made by sound may only be made in compliance with the requirements of the International Regulations for Preventing Collisions at Sea, Rules 34 and 35, accepting that sound signals "G" and "Z" may continue to be used by fishing vessels fishing in close proximity to other fishing vessels.

2. Signals "K" and "S" have special meanings as landing signals for small boats with crews or persons in distress. (International Convention for the Safety of Life at Sea, 1960, Chapter V, Regulation 16).

Chapter XII Single-letter signals with complements

May be made by any method of signalling

A	with three numerals	AZIMUTH or BEARING
C	with three numerals	COURSE
D	with two, four or six numerals	DATE
G	with four or five numerals	LONGITUDE (the last two numerals denote minutes and the rest degrees)
K	with one numeral	I wish to COMMUNICATE with you by . . . (complements table I)
L	with four numerals	LATITUDE (the first two denote degrees and the rest minutes)
R	with one or more numerals	DISTANCE in nautical miles
S	with one or more numerals	SPEED in knots
T	with four numerals	LOCAL TIME (the first two denote hours and the rest minutes
V	with one or more numerals	SPEED in kilometres per hour
Z	with four numerals	GMT (the first two denote hours and the rest minutes)

AZIMUTH or BEARING	A with three numerals
COMMUNICATE, I wish to communicate with you by . . . (complements table I)	K with one numeral
COURSE	C with three numerals
DATE	D with two, four or six numerals
DISTANCE in nautical miles	R with one or more numerals
GMT (the first two denote hours and the rest minutes)	Z with four numerals
LATITUDE (the first two denote degrees and the rest minutes)	L with four numerals
LONGITUDE (the last two numerals denote minutes and the rest degrees)	G with four or five numerals
LOCAL TIME (the first two denote hours and the rest minutes)	T with four numerals
SPEED in kilometres per hour	V with one or more numerals
SPEED in knots	S with one or more numerals

Chapter XIII Single-letter signals between ice-breaker and assisted vessels

The following single-letter signals, when made between an ice-breaker and assisted vessels, have only the significations given in this table and are only to be made by sound, visual or radiotelephony signals.

WM Ice-breaker support is now commencing. Use special ice-breaker support signals and keep continuous watch for sound, visual or radiotelephony signals.

WO Ice-breaker support is finished. Proceed to your destination.

Code letters or figures	Ice-breaker	Assisted vessel(s)
A ·–	Go ahead (proceed along the ice channel).	I am going ahead (I am proceeding along the ice channel).
G ––·	I am going ahead; follow me.	I am going ahead; I am following you.
J ·–––	Do not follow me (proceed along the ice channel).	I will not follow you (I will proceed along the ice channel).
P ·––·	Slow down.	I am slowing down.
N –·	Stop your engines.	I am stopping my engines.
H ····	Reverse your engines.	Reverse your engines.
L ·–··	You should stop your vessel instantly.	I am stopping my vessel.
4 ····–	Stop. I am ice-bound.	Stop. I am ice-bound.
Q ––·–	Shorten the distance between vessels.	I am shortening the distance.
B –···	Increase the distance between vessels.	I am increasing the distance.
5 ·····	Attention.	Attention.
Y –·––	Be ready to take (or cast off) the tow line.	I am ready to take (or cast off) the tow line.

Notes:
1. The signal "K" (–·–) by sound or light may be used by an ice-breaker to remind ships of their obligation to listen continuously on their radio.
2. If more than one vessel is assisted, the distances between vessels should be as constant as possible; watch speed of your own vessel and vessel ahead. Should speed of your own vessel go down, give attention signal to the following vessel.
3. The use of these signals does not relieve any vessel from complying with the International Regulations for Preventing Collisions at Sea.
4. ···–·· Stop your headway (given only to a ship in an ice-channel ahead of and approaching or going away from ice-breaker).   I am stopping headway.

This signal should not be made by radiotelephone.

Single-letter signals between ice-breaker and assisted vessels

Single-letter signals which may be used during ice-breaking operations:

*E   .   I am altering my course to starboard.

*I   . .   I am altering my course to port.

*S   . . .   I am operating astern propulsion.

M   – –   My vessel is stopped and making no way through the water.

Notes:
1. Signals of letters marked*, when made by sound, may only be made in compliance with the requirements of the International Regulations for Preventing Collisions at Sea.

2. Additional signals for ice-breaking support can be found on pages 86–87.

Chapter XIV Identification of medical transport in armed conflict and permanent identification of rescue craft*

1 SHAPE, COLOUR AND POSITIONING OF EMBLEMS FOR MEDICAL TRANSPORTS

1.1 The following emblems can be used separately or together to show that a vessel is protected as a medical transports under the Geneva Convention.

1.2 The emblem, positioned on the vessel's sides, bow, stern and deck, shall be painted dark red on a white background.

1.3 The emblem shall be as large as possible.

1.3.1 On the vessel's sides the emblem shall extend from the waterline to the top of the ship's hull.

1.3.2 The emblems on the vessel's bow and stern must, if necessary, be painted on a wooden structure so as to be clearly visible to other vessels ahead or astern.

1.3.3 The deck emblem must be as clear of the vessel's equipment as possible to be clearly visible from aircraft.

1.4 In order to provide the desired contrast for infra-red film or instruments, the red emblem must be painted on top of a black primer paint.

1.5 Emblems may also be made of materials which make them recognizable by technical means of detecting.

2 ILLUMINATION
2.1 At night and in restricted visibility the emblems shall be illuminated or lit.

2.2 At night and in restricted visibility all deck and overside lights must be fully lit to indicate that the vessel is engaged in medical operations.

3. PERSONAL EQUIPMENT
3.1 Subject to the instructions of the competent authority, medical and religious personnel carrying out their duties in the battle area shall, as far as possible, wear headgear and clothing bearing the distinctive emblem.

---

* In accordance with Article 27 of the second Geneva Convention of 12 August 1949 this chapter also applies to coastal rescue craft.

4 FLASHING BLUE LIGHT FOR MEDICAL TRANSPORTS

4.1 A vessel engaged in medical operations shall exhibit one or more all-round flashing blue lights of the colour prescribed in paragraph 4.4.

4.2 The visibility of the lights shall be as high as possible and not less than 3 nautical miles in accordance with Annex 1 to the International Regulations for Preventing Collisions at Sea, 1972.

4.3 The light or lights shall be exhibited as high above the hull as practical and in such a way that at least one light shall be visible from any direction.

4.4 The recommended blue colour is obtained by using, as trichromatic co-ordinates:

green boundary $y = 0.065 + 0.805x$
white boundary $y = 0.400 - x$
purple boundary $x = 0.133 + 0.600y$

4.5 The frequency of the flashing light shall be between 60 and 100 flashes per minute.

5 RADAR TRANSPONDERS

5.1 It should be possible for medical transports to be identified by other vessels equipped with radar by signals from a radar transponder fitted on the medical transport.

5.2 The signal from the medical transport transponder shall consist of the group YYY, in accordance with Article 40 of the Radio Regulations followed by the call sign of the ship or other recognized means of identification.

6 UNDERWATER ACOUSTIC SIGNALS

6.1 It should be possible for medical transports to be identified by submarines by appropriate underwater signals transmitted by the medical transports.

6.2 The underwater signal shall consist of the call sign of the ship preceded by the single group YYY transmitted in morse on an appropriate acoustic frequency, e.g. 5kHz.

7 RESCUE CRAFT CARRIED BY MEDICAL TRANSPORTS

7.1 Every rescue craft should be equipped with a mast on which a red cross flag measuring about 2 × 2 metres can be hoisted.

8 FLASHING BLUE LIGHT FOR MEDICAL AIRCRAFT

8.1 The light signal, consisting of a flashing blue light, is established for the use of medical aircraft to signal their identity. No other aircraft shall use this signal. The recommended flashing rate of the blue light is between sixty and one hundred flashes per minute.

8.2 Medical aircraft should be equipped with such lights as may be necessary to make the light signal visible in as many directions as possible.

General Section

Part I Distress—emergency

Abandon

*AA Repeat all after . . .

*AB Repeat all before . . .

AC I am abandoning my vessel.

AD I am abandoning my vessel which has suffered a nuclear accident and is a possible source of radiation danger.

AE I must abandon my vessel.
    AE 1 I (or crew of vessel indicated) wish to abandon my (or their) vessel, but have not the means.
    AE 2 I shall abandon my vessel unless you will remain by me, ready to assist.

AF I do not intend to abandon my vessel.
    AF 1 Do you intend to abandon your vessel?

AG You should abandon your vessel as quickly as possible.

AH You should not abandon your vessel.

AI Vessel (indicated by position and/or name or identity signal if necessary) will have to be abandoned.

Accident – doctor – injured/sick

Accident

AJ I have had a serious nuclear accident and you should approach with caution.

AK I have had nuclear accident on board.

I am abandoning my vessel which has suffered a nuclear accident and is a possible source of radiation danger.	AD
I am proceeding to the position of accident.	SB
I am proceeding to the position of accident at full speed. Expect to arrive at time indicated.	FE
Are you proceeding to the position of accident? If so, when do you expect to arrive?	FE 1
You should steer course . . . (or follow me) to reach position of accident.	FL
I am circling over the area of accident.	BJ
An aircraft is circling over the area of accident.	BJ 1

*Procedural signals for repetition.

Position of accident (or survival craft) is marked.	FJ
Position of accident (or survival craft) is marked by flame or smoke float.	FJ 1
Position of accident (or survival craft) is marked by sea marker.	FJ 2
Position of accident (or survival craft) is marked by sea marker dye.	FJ 3
Position of accident (or survival craft) is marked by radio beacon.	FJ 4
Position of accident (or survival craft) is marked by wreckage.	FJ 5
Is position of accident (or survival craft) marked?	FK
I have searched area of accident but have found no trace of derelict or survivors.	GC 2
Man overboard. Please take action to pick him up (position to be indicated if necessary).	GW

Doctor

AL I have a doctor on board.

AM Have you a doctor?

AN I need a doctor.
  AN 1 I need a doctor; I have severe burns.
  AN 2 I need a doctor; I have radiation casualties.

I require a helicopter urgently, with a doctor.	BR 2
Helicopter is coming to you now (or at time indicated) with a doctor.	BT 2

Injured/sick

AO Number of injured and/or dead not yet known.
  AO 1 How many injured?
  AO 2 How many dead?

AP I have . . . (number) casualties.

AQ I have injured/sick person (or number of persons indicated) to be taken off urgently.

I cannot alight but I can lift injured/sick person.	AZ 1
You cannot alight on the deck; can you lift injured/sick person?	BA 2
I require a helicopter urgently to pick up injured/sick person.	BR 3

	You should send a helicopter/boat with a stretcher.	BS
	A helicopter/boat is coming to take injured/sick.	BU

AT You should send injured/sick persons to me.

Aircraft—helicopter

Alight—landing

AU I am forced to alight near you (or in position indicated).

AV I am alighting (in position indicated if necessary) to pick up crew of vessel/aircraft.

AW Aircraft should endeavour to alight where flag is waved or light is shown.

AX You should train your searchlight nearly vertical on a cloud, intermittently if possible, and, if my aircraft is seen, deflect the beam up wind and on the water to facilitate my landing.

    AX 1 Shall I train my searchlight nearly vertical on a cloud, intermittently if possible, and, if your aircraft is seen, deflect the beam up wind and on the water to facilitate your landing?

AY I will alight on your deck; (you should steer course . . . speed . . . knots).

AZ I cannot alight but I can lift crew.
    AZ 1 I cannot alight but I can lift injured/sick person.

BA You cannot alight on the deck.
    BA 1 You cannot alight on the deck; can you lift crew?
    BA 2 You cannot alight on the deck; can you lift injured/sick person?

BB You may alight on my deck.
    BB 1 You may alight on my deck; I am ready to receive you forward.
    BB 2 You may alight on my deck; I am ready to receive you amidship.
    BB 3 You may alight on my deck; I am ready to receive you aft.
    BB 4 You may alight on my deck but I am not yet ready to receive you.

Communications

BC I have established communications with the aircraft in distress on 2182 kHz
    BC 1 Can you communicate with the aircraft?

BD I have established communications with the aircraft in distress on . . . kHz

BE I have established communications with the aircraft in distress on . . . MHz

Ditched—disabled—afloat

BF Aircraft is ditched in position indicated and requires immediate assistance.

    I sighted disabled aircraft in lat . . . long . . . at time indicated.   DS

BG Aircraft is still afloat.

Flying

BH I sighted an aircraft at time indicated in lat . . . long . . . flying on course . . .
    BH 1 Aircraft was flying at high altitude.
    BH 2 Aircraft was flying at low altitude.

BI I am flying to likely position of vessel in distress.
    BI 1 I am flying at low altitude near the vessel.

BJ I am circling over the area of accident.
    BJ 1 An aircraft is circling over the area of accident.

BK You are overhead.
    BK 1 Am I overhead?

BL I am having engine trouble but am continuing flight.

Parachute

BM You should parachute object to windward. Mark it by smoke or light signal.
    BM 1 I am going to parachute object to windward, marking it by smoke or light signal.
    BM 2 I am going to parachute equipment.
    BM 3 Inflatable raft will be dropped to windward by parachute.

*BN Repeat all between . . . and . . .

BO We are going to jump by parachute.

Search—assistance

BP Aircraft is coming to participate in search. Expected arrive over the area of accident at time indicated.

    The search area of the aircraft is between lat . . . and . . ., and   FU
    long . . . and . . .

*Procedural signal for repetition.

Search by aircraft/helicopter will be discontinued because of unfavourable conditions.   FV

SAR aircraft is coming to your assistance.   CP 1

Speed

BQ   The speed of my aircraft in relation to the surface of the earth is . . . (knots or kilometres per hour).
      BQ 1   What is the speed of your aircraft in relation to the surface of the earth?

Helicopter

BR   I require a helicopter urgently.
      BR 1   I require a helicopter urgently to pick up persons.
      BR 2   I require a helicopter urgently with a doctor.
      BR 3   I require a helicopter urgently to pick up injured/sick person.
      BR 4   I require a helicopter urgently with inflatable raft.

BS   You should send a helicopter/boat with stretcher.

BT   Helicopter is coming to you now (or at time indicated).
      BT 1   Helicopter is coming to you now (or at time indicated) to pick up persons.
      BT 2   Helicopter is coming to you now (or at time indicated) with a doctor.
      BT 3   Helicopter is coming to you now (or at time indicated) to pick up injured/sick person.
      BT 4   Helicopter is coming to you now (or at time indicated) with inflatable raft.

BU   A helicopter/boat is coming to take injured/sick.

BV   I cannot send a helicopter.

BW   The magnetic course for you to steer towards me (or vessel or position indicated) is . . . (at time indicated).

BX   The magnetic course for the helicopter to regain its base is . . .

BY   Will you indicate the magnetic course for me to steer towards you (or vessel or position indicated)?

BZ   Your magnetic bearing from me (or from vessel or position indicated) is . . . (at time indicated).

CA   What is my magnetic bearing from you (or from vessel or position indicated)?

Assistance

Required

| | I am in distress and require immediate assistance. | NC |

- CB I require immediate assistance.
  - CB 1 I require immediate assistance; I have a dangerous list.
  - CB 2 I require immediate assistance; I have damaged steering gear.
  - CB 3 I require immediate assistance; I have a serious disturbance on board.
  - CB 4 I require immediate assistance; I am aground.
  - CB 5 I require immediate assistance; I am drifting.
  - CB 6 I require immediate assistance; I am on fire.
  - CB 7 I require immediate assistance; I have sprung a leak.
  - CB 8 I require immediate assistance; propeller shaft is broken.

- CC I am (or vessel indicated is) in distress in lat . . . long . . . (or bearing . . . from place indicated, distance . . .) and require immediate assistance (Complements table II, if required).

| | I require assistance. | V |

- CD I require assistance in the nature of . . . (Complements table II)

| | I require medical assistance. | W |
| | I request assistance from fishery protection (or fishery assistance) vessel. | TY |

- CE I will attempt to obtain for you the assistance required.

| | Aircraft is ditched in position indicated and requires immediate assistance. | BF |

- CF Signals from vessel/aircraft requesting assistance are coming from bearing . . . from me (lat . . . long . . . if necessary).

- CG Stand by to assist me (or vessel indicated).
  - CG 1 I will stand by to assist you (or vessel indicated).

| | Survivors are in bad condition. Medical assistance is urgently required. | HM |

- CH Vessel indicated is reported as requiring assistance in lat . . . long . . . (or bearing . . . from place indicated, distance . . .).
  - CH 1 Lightvessel (or lighthouse) indicated requires assistance.
  - CH 2 Space ship is down in lat . . . long . . . and requires immediate assistance.

- CI Vessel aground in lat . . . long . . . requires assistance.

CJ Do you require assistance?
    CJ 1 Do you require immediate assistance?
    CJ 2 Do you require any further assistance?
    CJ 3 What assistance do you require?
    CJ 4 Can you proceed without assistance?

Not required—declined

CK Assistance is not (or is no longer) required by me (or vessel indicated).

CL I offered assistance but it was declined.

Given—not given

CM One or more vessels are assisting the vessel in distress.
    CM 1 Vessel/aircraft reported in distress is receiving assistance.

CN You should give all possible assistance.
    CN 1 You should give immediate assistance to pick up survivors.
    CN 2 You should send survival craft to assist vessel indicated.

CO Assistance cannot be given to you (or vessel/aircraft indicated).
    CO 1 I cannot give the assistance required.

Proceding to assistance

CP I am (or vessel indicated is) proceeding to your assistance.
    CP 1 SAR aircraft is coming to your assistance.

*CQ Call for unknown station(s) or general call to all stations.

CR I am proceeding to the assistance of vessel/aircraft in distress (lat . . . long . . .).

*CS What is the name or identity signal of your vessel (or station)?

CT I (or vessel indicated) expect to reach you at time indicated.

CU Assistance will come at time indicated.
    CU 1 I can assist you.

CV I am unable to give assistance.
    CV 1 Will you go to the assistance of vessel indicated (in lat . . . long . . .)?
    CV 2 May I assist you?
    CV 3 Can you assist me (or vessel indicated)?
    CV 4 Can you assist?

Can you offer assistance? (Complements table II)   TZ

*Procedural signals.

I shall abandon my vessel unless you will remain by me, ready to assist.	AE 2
I cannot get the fire under control without assistance.	IX 1
I can get the fire under control without assistance.	IY
Can you get the fire under control without assistance?	IY 1
I have placed the collision mat. I can proceed without assistance.	KA 1
I cannot take you (or vessel indicated) in tow, but I will report you and ask for immediate assistance.	KN 1
I cannot steer without assistance.	PK

Boats—rafts

CW Boat/raft is on board.
    CW 1 Boat/raft is safe.
    CW 2 Boat/raft is in sight.
    CW 3 Boat/raft is adrift.
    CW 4 Boat/raft is aground.
    CW 5 Boat/raft is alongside.
    CW 6 Boat/raft is damaged.
    CW 7 Boat/raft has sunk.
    CW 8 Boat/raft has capsized.

CX Boats cannot be used.
    CX 1 Boats cannot be used because of prevailing weather conditions.
    CX 2 Boats cannot be used on the starboard side because of list.
    CX 3 Boats cannot be used on the port side because of list.
    CX 4 Boats cannot be used to disembark people.
    CX 5 Boats cannot be used to get alongside.
    CX 6 Boats cannot be used to reach you.
    CX 7 I cannot send a boat.

CY Boat(s) is (are) coming to you.
    CY 1 Boat/raft is making for the shore.
    CY 2 Boat/raft has reached the shore.

CZ You should make a lee for the boat(s)/raft(s).
    CZ 1 You should discharge oil to smooth sea.

DA Boat(s)/raft(s) should approach vessel as near as possible to take off persons.

A boat/helicopter is coming to take injured/sick.	BU

DB Veer a boat or raft on a line.

DC Boat should endeavour to land where flag is waved or light is shown.

DD Boats are not allowed to come alongside.
    DD 1 Boats are not allowed to land (after time indicated).

*DE From . . .

Available

DF I have . . . (number) serviceable boats.

DG I have a motor boat [or . . . (number) motor boats].

DH I have no boat/raft.
    DH 1 I have no motor boat.
    DH 2 Have you any boats with radiotelegraph installation or portable radio equipment?
    DH 3 How many serviceable motor boats have you?
    DH 4 How many serviceable boats have you?

Required

DI I require boats for . . . (number) persons.

DJ Do you require a boat?

Send

DK You should send all available boats/rafts.
    DK 1 You should send back my boat.
    DK 2 Can you send a boat?

You should send a boat/helicopter with stretcher.	BS
You should send survival craft to assist vessel indicated.	CN 2
You should stop, or heave to; I am going to send a boat.	SQ 2

DL I can send a boat.
    DL 1 I am sending a boat.

I cannot send a boat.	CX 7

Search

DM You should search for the boat(s)/raft(s).

DN I have found the boat/raft.
    DN 1 Have you seen or heard anything of the boat/raft?

---

*Procedural signal used to precede the name or identity signal of the calling station.

DO	Look out for boat/raft in bearing . . . distance . . . from me (or from position indicated).
DP	There is a boat/raft in bearing . . . distance . . . from me (or from position indicated).
DQ	An empty boat/raft has been sighted in lat . . . long . . . (or bearing . . . from place indicated, distance . . .).

Disabled—drifting—sinking

Disabled

DR	Have you sighted disabled vessel/aircraft in approximate lat . . . long . . . ?	
DS	I sighted disabled aircraft in lat . . . long . . . at time indicated.	
DT	I sighted disabled vessel in lat . . . long . . . at time indicated. DT 1 I sighted disabled vessel in lat . . . long . . . at time indicated, apparently without a radio.	
	I am disabled; communicate with me.	F

Drifting

DU	I am drifting at . . . (number) knots, towards . . . degrees.	
DV	I am drifting DV 1 I am adrift.	
DW	Vessel (name or identity signal) is drifting near lat . . . long	
	I require immediate assistance; I am drifting.	CB 5
	I am (or vessel indicated is) breaking adrift.	RC
	I have broken adrift.	RC 1

Sinking

DX	I am sinking (lat . . . long . . . if necessary).
DY	Vessel (name or identity signal) has sunk in lat . . . long . . . DY 1 Did you see vessel sink? DY 2 Where did vessel sink? DY 3 Is it confirmed that vessel (name or identity signal) has sunk? DY 4 What is the depth of water where vessel sunk?

Distress

Vessel/aircraft in distress

	I am in distress and require immediate assistance.	NC

- DZ Vessel (or aircraft) indicated appears to be in distress.
  - DZ 1 Is vessel (or aircraft) indicated in distress?
  - DZ 2 What is the name (or identity signal) of vessel in distress?

- EA Have you sighted or heard of a vessel in distress? (Approximate position lat . . . long . . . or bearing . . . from place indicated, distance . . .).
  - EA 1 Have you any news of vessel/aircraft reported missing or in distress in this area?

I am (or vessel indicated is) in distress in lat . . . long . . . (or bearing . . . from place indicated, distance . . .) and require immediate assistance (Complements table II if required).	CC

- EB There is a vessel (or aircraft) in distress in lat . . . long . . . (or bearing . . . distance . . . from me, or Complements table III).

- EC A vessel which has suffered a nuclear accident is in distress in lat . . . long . . .

Distress signals

- *ED Your distress signals are understood.
  - ED 1 Your distress signals are understood; the nearest life-saving station is being informed.

- EF SOS/MAYDAY has been cancelled.
  - EF 1 Has the SOS/MAYDAY been cancelled?

I have intercepted SOS/MAYDAY from vessel (name or identity signal) (or aircraft) in position lat . . . long . . . at time indicated.	FF

- EG Did you hear SOS/MAYDAY given at time indicated?
  - EG 1 Will you listen on 2182 kHz for signals of emergency position-indicating radio beacons?
  - EG 2 I am listening on 2182 kHz for signals of emergency position-indicating radio beacons.
  - EG 3 Have you received the signal of an emergency position-indicating radio beacon on 2182 kHz?
  - EG 4 I have received the signal of an emergency position-indicating radio beacon on 2182 kHz
  - EG 5 Will you listen on . . . MHz for signals of emergency position-indicating radio beacons?

*Reference is made to signals prescribed by the International Convention for the Safety of Life at Sea, 1960 (Regulation 16(a), Chapter V) as replies from life-saving stations or maritime rescue units to distress signals made by a ship or person.

EG 6   I am listening on . . . MHz for signals of emergency position-indicating radio beacons.

EG 7   Have you received the signal of an emergency position-indicating radio beacon on . . . MHz?

EG 8   I have received the signal of an emergency position-indicating radio beacon on . . . MHz.

EJ   I have received distress signal transmitted by coast station indicated.

    EJ 1   Have you received distress signal transmitted by coast station indicated?

EK   I have sighted distress signal in lat . . . long . . .

    EK 1   An explosion was seen or heard (position or direction and time to be indicated).

    EK 2   Have you heard or seen distress signal from survival craft?

Position of distress

EL   Repeat the distress position.

    EL 1   What is the position of vessel in distress?

Position given with SOS/MAYDAY from vessel (or aircraft) was lat . . . long . . . (or bearing . . . from place indicated, distance . . .).	FG
What was the position given with SOS/MAYDAY from vessel (or aircraft)?	FG 1
Position given with SOS/MAYDAY is wrong. The correct position is lat . . . long . . .	FH
Position given with SOS/MAYDAY by vessel is wrong. I have her bearing by radio direction-finder and can exchange bearings with any other vessel.	FI
Survival craft are believed to be in the vicinity of lat . . . long . . .	GI

EM   Are there other vessels/aircraft in the vicinity of vessel/aircraft in distress?

Contact or locate

EN   You should try to contact vessel/aircraft in distress.

EO   I am unable to locate vessel/aircraft in distress because of poor visibility.

EP   I have lost sight of you.

I have located (or found) wreckage from the vessel/aircraft in distress (position to be indicated if necessary by lat . . . and long . . . or by bearing . . . from specified place, and distance . . .).	GL

EQ	I expect to be at the position of vessel/aircraft in distress at time indicated. EQ 1 Indicate estimated time of your arrival at position of vessel/aircraft in distress.	
	I am flying to likely position of vessel in distress.	BI
	One or more vessels are assisting the vessel in distress.	CM
	Vessel/aircraft reported in distress is receiving assistance.	CM 1
	I am proceeding to the assistance of vessel/aircraft in distress in distress in lat . . . long . . .	CR
	I have found vessel/aircraft in distress in lat . . . long . . .	GF

Position

ER You should indicate your position at time indicated.

ET My position at time indicated was lat . . . long . . .

EU My present position is lat . . . long . . . (or bearing . . . from place indicated, distance . . .).
    EU 1 What is your present position?

EV My present position, course and speed are lat . . . long . . . , . . . , knots . . .
    EV 1 What are your present position, course and speed?

EW My position is ascertained by dead reckoning.
    EW 1 My position is ascertained by visual bearings.
    EW 2 My position is ascertained by astronomical observations.
    EW 3 My position is ascertained by radio beacons.
    EW 4 My position is ascertained by radar.
    EW 5 My position is ascertained by an electronic position-fixing system.

EX My position is doubtful.

EY I am confident as to my position.
    EY 1 Are you confident as to your position?

EZ Your position according to bearings taken by radio direction-finder stations which I control is lat . . . long . . . (at time indicated).
    EZ 1 Will you give me my position according to bearings taken by radio direction-finder stations which you control?

FA Will you give me my position?

FB Will vessels in my immediate vicinity (or in the vicinity of lat . . . long . . .) please indicate position, course and speed.

Position of distress

FC   You should indicate your position by visual or sound signals.
    FC 1   You should indicate your position by rockets or flares.
    FC 2   You should indicate your position by visual signals.
    FC 3   You should indicate your position by sound signals.
    FC 4   You should indicate your position by searchlight.
    FC 5   You should indicate your position by smoke signal.

FD   My position is indicated by visual or sound signals.
    FD 1   My position is indicated by rockets or flares.
    FD 2   My position is indicated by visual signals.
    FD 3   My position is indicated by sound signals.
    FD 4   My position is indicated by searchlight.
    FD 5   My position is indicated by smoke signal.

I expect to be at the position of vessel/aircraft in distress at time indicated.	EQ
Indicate estimated time of your arrival at position of vessel/aircraft in distress.	EQ 1
Position given with SOS/MAYDAY from vessel (or aircraft) was lat . . . long . . . (or bearing . . . from place indicated, distance . . .).	FG
What was position given with SOS/MAYDAY from vessel (or aircraft)?	FG 1
Position given with SOS/MAYDAY is wrong. The correct position is lat . . . long . . .	FH
Position given with SOS/MAYDAY by vessel is wrong. I have her bearing by radio direction-finder and can exchange bearings with any other vessel.	FI
Position of accident (or survival craft) is marked.	FJ
Position of accident (or survival craft) is marked by flame (or smoke float).	FJ 1
Position of accident (or survival craft) is marked by sea marker.	FJ 2
Position of accident (or survival craft) is marked by sea marker dye.	FJ 3
Position of accident (or survival craft) is marked by radio beacon.	FJ 4
Position of accident (or survival craft) is marked by wreckage.	FJ 5
Is position of accident (or survival craft) marked?	FK
You should transmit your identification and series of long dashes or your carrier frequency to home vessel (or aircraft) to your position.	FQ
Shall I home vessel (or aircraft) to my position?	FQ 1
You should indicate position of survivors by throwing pyrotechnic signals.	HT

Search and rescue

Assistance, proceeding to

	I am proceeding to the assistance of vessel/aircraft in distress (lat . . . long . . .).	CR
FE	I am proceeding to the position of accident at full speed. Expect to arrive at time indicated.	
	FE 1 Are you proceeding to the position of accident? If so, when do you expect to arrive?	
	I am unable to give assistance	CV
	Can you assist?	CV 4

Position of distress or accident

FF I have intercepted SOS/MAYDAY from vessel (name or identity signal) (or aircraft) in position lat . . . long . . . at time indicated.
     FF 1 I have intercepted SOS/MAYDAY from vessel (name or identity signal) (or aircraft) in position lat . . . long . . . at time indicated; I have heard nothing since.

FG Position given with SOS/MAYDAY from vessel (or aircraft) was lat . . . long . . . (or bearing . . . from place indicated, distance . . .).
     FG 1 What was position given with SOS/MAYDAY from vessel (or aircraft)?

FH Position given with SOS/MAYDAY is wrong. The correct position is lat . . . long . . .

FI Position given with SOS/MAYDAY by vessel is wrong. I have bearing by radio direction-finder and can exchange bearings with any other vessel.

FJ Position of accident (or survival craft) is marked.
     FJ 1 Position of accident (or survival craft) is marked by flame or smoke float.
     FJ 2 Position of accident (or survival craft) is marked by sea marker.
     FJ 3 Position of accident (or survival craft) is marked by sea marker dye.
     FJ 4 Position of accident (or survival craft) is marked by radio beacon.
     FJ 5 Position of accident (or survival craft) is marked by wreckage.

FK Is position of accident (or survival craft) marked?

Information – instructions

FL	You should steer course . . . (or follow me) to reach position of accident.	
	Course to reach me is . . .	MF
	What is the course to reach you?	MF 1
FM	Visual contact with vessel is not continuous.	
FN	I have lost all contact with vessel.	
	I have lost sight of you.	EP
FO	I will keep close to you.	
	FO 1 I will keep close to you during the night.	
FP	Estimated set and drift of survival craft is . . . degrees and . . . knots.	
	FP 1 What is the estimated set and drift of survival craft?	
FQ	You should transmit your identification and series of long dashes or your carrier frequency to home vessel (or aircraft) to your position.	
	FQ 1 Shall I home vessel (or aircraft) to my position?	

Search

- FR I am (or vessel indicated is) in charge of co-ordinating search.
    - *FR 1 Carry out search pattern . . . starting at . . . hours. Initial course . . . search speed . . . knots.
    - *FR 2 Carry out radar search, ships proceeding in loose line abreast at intervals between ships . . . miles. Initial course . . . search speed . . . knots.
    - *FR 3 Vessel indicated (call sign or identity signal) is allocated track number . . .
    - *FR 4 Vessels(s) indicated adjust interval between ships to . . . miles.
    - *FR 5 Adjust track spacing to . . . miles.
    - *FR 6 Search speed will now be . . . knots.
    - *FR 7 Alter course as necessary to next leg of track now (or at time indicated).
- FS Please take charge of search in sector stretching between bearings . . . and . . . from vessel in distress.
- FT Please take charge of search in sector between lat . . . and . . . , and long . . . and . . .
- FU The search area of the aircraft is between lat . . . and . . . , and long . . . and . . .
- FV Search by aircraft/helicopter will be discontinued because of unfavourable conditions.
- FW You should search in the vicinity of lat . . . long . . .
- FX Shall I search in the vicinity of lat . . . long . . . ?
- FY I am in the search area.
    - FY 1 Are you in the search area?

	Aircraft is coming to participate in search. Expected arrive over the area of accident at time indicated.	BP
FZ	You should continue search according to instructions and until further notice.	
	FZ 1 I am continuing to search.	

*These signals are intended for use in connection with the Merchant Ship Search and Rescue Manual (MERSAR).

FZ 2   Are you continuing to search?
FZ 3   Do you want me to continue to search?

GA   I cannot continue to search.

GB   You should stop search and return to base or continue your voyage.

Results of search

GC   Report results of search.
GC 1 Results of search negative. I am continuing to search.
GC 2 I have searched area of accident but have found no trace of derelict or survivors.
GC 3 I have noted patches of oil at likely position of accident.

GD   Vessel/aircraft missing or being looked for has not been heard of since.
GD 1 Have you anything to report on vessel/aircraft missing or being looked for?
GD 2 Have you seen wreckage (or derelict)?

GE   Vessel/aircraft has been located at lat . . . long . . .

GF   I have found vessel/aircraft in distress in lat . . . long . . .

GG   Vessel/aircraft was last reported at time indicated in lat . . . long . . . steering course . . .

GH   I have sighted survival craft in lat . . . long . . . (or bearing . . . distance . . . from me).

GI   Survival craft are believed to be in the vicinity of lat . . . long . . .

GJ   Wreckage is reported in lat . . . long . . .
GJ 1 Wreckage is reported in lat . . . long . . . No survivors appear to be in the vicinity.

GK   Aircraft wreckage is found in lat . . . long . . .

GL   I have located (or found) wreckage from the vessel/aircraft in distress (position to be indicated if necessary by lat . . . and long . . . or by bearing . . . from specified place and distance . . .).

Rescue

GM   I cannot save my vessel.
GM 1 I cannot save my vessel; keep as close as possible.

GN   You should take off persons.
GN 1 I wish some persons taken off. Skeleton crew will remain on board.
GN 2 I will take off persons.
GN 3 Can you take off persons?

GO	I cannot take off persons.	
GP	You should proceed to the rescue of vessel (or ditched aircraft) in lat . . . long . . .	
GQ	I cannot proceed to the rescue owing to weather. You should do all you can.	
GR	Vessel coming to your rescue (or to the rescue of vessel or aircraft indicated) is steering course . . ., speed . . . knots. GR 1 You should indicate course and speed of vessel coming to my rescue (or to the rescue of vessel or aircraft indicated).	
GS	I will attempt rescue with whip and breeches buoy.	
*GT	I will endeavour to connect with line-throwing apparatus. GT 1 Look out for rocket-line.	
GU	It is not safe to fire a rocket.	
GV	You should endeavour to send me a line. GV 1 Have you a line-throwing apparatus? GV 2 Can you connect with line-throwing apparatus? GV 3 I have not a line-throwing apparatus.	
GW	Man overboard. Please take action to pick him up (position to be indicated if necessary).	
	Man overboard.	O

Results of rescue

GX	Report results of rescue. GX 1 What have you (or rescue vessel/aircraft) picked up?
GY	I (or rescue vessel/aircraft) have picked up wreckage.
GZ	All persons saved. GZ 1 All persons lost.
HA	I (or rescue vessel/aircraft) have rescued . . . (number) injured persons.
HB	I (or rescue vessel/aircraft) have rescued . . . (number) survivors.
HC	I (or rescue vessel/aircraft) have picked up . . . (number) bodies.
HD	Can I transfer rescued persons to you?

*Reference is made to signals prescribed by the International Convention for the Safety of Life at Sea, 1974 (Regulation 16 Chapter V) in connexion with the use of shore life-saving apparatus.

Survivors

HF I have located survivors in water, lat . . . long . . . (or bearing . . . from place indicated, distance . . .).

HG I have located survivors in survival craft lat . . . long . . . (or bearing . . . from place indicated, distance . . .).

HJ I have located survivors on drifting ice, lat . . . long . . .

HK I have located bodies in lat . . . long . . . (or bearing . . . from place indicated, distance . . .).

HL Survivors not yet located.
    HL 1 I am still looking for survivors.
    HL 2 Have you located survivors? If so, in what position?

HM Survivors are in bad condition. Medical assistance is urgently required.
    HM 1 Survivors are in bad condition.
    HM 2 Survivors are in good condition.
    HM 3 Condition of survivors not ascertained.
    HM 4 What is condition of survivors?

HN You should proceed to lat . . . long . . . to pick up survivors.

HO Pick up survivors from drifting ice, lat . . . long . . .
    HO 1 Pick up survivors from sinking vessel/aircraft.

HP Survivors have not yet been picked up.
    HP 1 Have survivors been picked up?

You should give immediate assistance to pick up survivors.   CN 1

HQ Transfer survivors to my vessel (or vessel indicated).
    HQ 1 Have you any survivors on board?

HR You should try to obtain from survivors all possible information.

HT You should indicate position of survivors by throwing pyrotechnic signals.

Part II Casualties—damages

Collision

HV Have you been in collision?

HW I have (or vessel indicated has) collided with surface craft.
    HW 1 I have (or vessel indicated has) collided with lightvessel.
    HW 2 I have (or vessel indicated has) collided with submarine.
    HW 3 I have (or vessel indicated has) collided with unknown vessel.
    HW 4 I have (or vessel indicated has) collided with underwater object.
    HW 5 I have (or vessel indicated has) collided with navigation buoy.
    HW 6 I have (or vessel indicated has) collided with iceberg.
    HW 7 I have (or vessel indicated has) collided with floating ice.

HX Have you received any damage in collision?
    HX 1 I have received serious damage above the water-line.
    HX 2 I have received serious damage below the water-line.
    HX 3 I have received minor damage above the water-line.
    HX 4 I have received minor damage below the water-line.

HY The vessel (name or identity signal) with which I have been in collision has sunk.
    HY 1 The vessel (name or identity signal) with which I have been in collision has resumed her voyage.
    HY 2 I do not know what has happened to the vessel with which I collided.
    HY 3 Has the vessel with which you have been in collision resumed her voyage?
    HY 4 What is the name (or identity signal) of the vessel with which you collided?
    HY 5 What is the name (or identity signal) of vessel which collided with me? My name (or identity signal) is . . .
    HY 6 Where is the vessel with which you collided?

HZ There has been a collision between vessels indicated (names or identity signals).

I urgently require a collision mat.	KA
I have placed the collision mat. I can proceed without assistance.	KA 1
Can you place the collision mat?	KA 2

Damages—repairs

IA I have received damage to stem.
    IA 1 I have received damage to stern frame.
    IA 2 I have received damage to side plate above water.
    IA 3 I have received damage to side plate below water.
    IA 4 I have received damage to bottom-plate.
    IA 5 I have received damage to boiler room.
    IA 6 I have received damage to engine room.
    IA 7 I have received damage to hatchways.

IA 8 I have received damage to steering gear.
IA 9 I have received damage to propellers.

IB What damage have you received?
- IB 1 My vessel is seriously damaged.
- IB 2 I have minor damage.
- IB 3 I have not received any damage.
- IB 4 The extent of the damage is still unknown.

Have you received any damage in collision?	HX
I have received serious damage above the water-line.	HX 1
I have received serious damage below the water-line.	HX 2
I have received minor damage above the water-line.	HX 3
I have received minor damage below the water-line.	HX 4

Can damage be repaired at sea?
- IC 1 Can damage be repaired at sea without assistance?
- IC 2 How long will it take you to repair damage?

ID Damage can be repaired at sea.
- ID 1 Damage can be repaired at sea without assistance.
- ID 2 Damage has been repaired.

IF Damage cannot be repaired at sea.
- IF 1 Damage cannot be repaired at sea without assistance.

IG Damage can be repaired in . . . (number) hrs.

IJ I will try to proceed by my own means but I request you to keep in contact with me by . . . (Complements table I).

IK I can proceed at . . . (number) knots.

IL I can only proceed at slow speed.
- IL 1 I can only proceed with one engine.
- IL 2 I am unable to proceed under my own power.
- IL 3 Are you in a condition to proceed?

IM I request to be escorted until further notice.

Propeller shaft is broken.	RO
My propeller is fouled by hawser or rope.	RO 1
I have lost my propeller.	RO 2

Diver—underwater operations

IN I require a diver.
- IN 1 I require a diver to clear propeller.
- IN 2 I require a diver to examine bottom.
- IN 3 I require a diver to place collision mat.
- IN 4 I require a diver to clear my anchor.

IO   I have no diver.

IP   A diver will be sent as soon as possible (or at time indicated).

IQ   Diver has been attacked by diver's disease and requires decompression chamber treatment.

*IR I am engaged in submarine survey work (underwater operations). Keep clear of me and go slow.

I have a diver down; keep well clear at slow speed.   A

Fire—explosion

Fire

IT   I am on fire.
    IT 1 I am on fire and have dangerous cargo on board; keep well clear of me.   J
    IT 2 Vessel (name or identity signal) is on fire.
    IT 3 Are you on fire?

IU   Vessel (name or identity signal) on fire is located at lat . . . long . . .

I require immediate assistance; I am on fire.   CB 6

IV   Where is the fire?
    IV 1 I am on fire in the engine room.
    IV 2 I am on fire in the boiler room.
    IV 3 I am on fire in hold or cargo.
    IV 4 I am on fire in passenger's or crew's quarters.
    IV 5 Oil is on fire.

IW   Fire is under control.

IX   Fire is gaining.
    IX 1 I cannot get the fire under control without assistance.
    IX 2 Fire has not been extinguished.

IY   I can get the fire under control without assistance.
    IY 1 Can you get the fire under control without assistance?

IZ   Fire has been extinguished.
    IZ 1 I am flooding compartment to extinguish fire.
    IZ 2 Is fire extinguished?

JA   I require fire-fighting appliances.
    JA 1 I require foam fire extinguishers.
    JA 2 I require CO_2 fire extinguishers.

---

*The use of this signal does not relieve any vessel from compliance with Rule 4(c) of the International Regulations for Preventing Collisions at Sea.

JA 3   I require carbon tetrachloride fire extinguishers.
JA 4   I require material for foam fire extinguishers.
JA 5   I require material for CO_2 fire extinguishers.
JA 6   I require material for carbon tetrachloride fire extinguishers.
JA 7   I require water pumps.

Explosion

JB   There is danger of explosion.

JC   There is no danger of explosion.
    JC 1   Is there any danger of explosion?

JD   Explosion has occurred in boiler.
    JD 1   Explosion has occurred in tank.
    JD 2   Explosion has occurred in cargo.
    JD 3   Further explosions are possible.
    JD 4   There is danger of toxic effects.

JE   Have you any casualties owing to explosion?

An explosion was seen or heard (position or direction and time to be indicated).   EK 1

Grounding—beaching—refloating

Grounding

JF   I am (or vessel indicated is) aground in lat ... long ... (also the following complements, if necessary):
    0   On rocky bottom.
    1   On soft bottom.
    2   Forward.
    3   Amidship.
    4   Aft.
    5   At high water forward.
    6   At high water amidship.
    7   At high water aft.
    8   Full length of vessel.
    9   Full length of vessel at high water.

JG   I am aground; I am in dangerous situation.

JH   I am aground; I am not in danger.

I require immediate assistance; I am aground.   CB 4

Vessel aground in lat ... long ... requires assistance.   CI

JI   Are you aground?
    JI 1   What was your draught when you went aground?
    JI 2   On what kind of ground have you gone aground?
    JI 3   At what state of tide did you go aground?
    JI 4   What part of your vessel is aground?

JJ My maximum draught when I went aground was ... (number) feet or metres.

JK The tide was high water when the vessel went aground.
    JK 1 The tide was half water when the vessel went aground.
    JK 2 The tide was low water when the vessel went aground.

JL You are running the risk of going aground.
    JL 1 You are running the risk of going aground; do not approach me from the starboard side.
    JL 2 You are running the risk of going aground; do not approach me from the port side.
    JL 3 You are running the risk of going aground; do not approach me from forward.
    JL 4 You are running the risk of going aground; do not approach me from aft.

JM You are running the risk of going aground at low water.

Beaching

JN You should beach the vessel in lat ... long ...
    JN 1 You should beach the vessel where flag is waved or light is shown.
    JN 2 I must beach the vessel.

Refloating

JO I am afloat.
    JO 1 I am afloat forward.
    JO 2 I am afloat aft.
    JO 3 I may be got afloat if prompt assistance is given.
    JO 4 Are you (or vessel indicated) still afloat?
    JO 5 When do you expect to be afloat?

JP I am jettisoning to refloat (the following complements should be used if required):
1 Cargo.
2 Bunkers.
3 Everything movable forward.
4 Everything movable aft.

JQ I cannot refloat without jettisoning (the following complements should be used if required):
1 Cargo.
2 Bunkers.
3 Everything movable forward.
4 Everything movable aft.

JR I expect (or vessel indicated expects) to refloat.
    JR 1 I expect (or vessel indicated expects) to refloat at time indicated.
    JR 2 I expect (or vessel indicated expects) to refloat in daylight.

JR 3 I expect(or vessel indicated expects) to refloat when tide rises.
JR 4 I expect (or vessel indicated expects) to refloat when visibility improves.
JR 5 I expect (or vessel indicated expects) to refloat when weather moderates.
JR 6 I expect (or vessel indicated expects) to refloat when draught is lightened.
JR 7 I expect (or vessel indicated expects) to refloat when tugs arrive.

JS Is it likely that you (or vessel indicated) will refloat?
JS 1 Is it likely that you (or vessel indicated) will refloat at time indicated?
JS 2 Is it likely that you (or vessel indicated) will refloat in daylight?
JS 3 Is it likely that you (or vessel indicated) will refloat when tide rises?
JS 4 Is it likely that you (or vessel indicated) will refloat when visibility improves?
JS 5 Is it likely that you (or vessel indicated) will refloat when weather moderates?
JS 6 Is it likely that you (or vessel indicated) will refloat when draught is lightened?
JS 7 Is it likely that you (or vessel indicated) will refloat when tugs arrive?

JT I can refloat if an anchor is laid out for me.
JT 1 I may refloat without assistance
JT 2 Will you assist me to refloat?

JU I cannot be refloated by any means now available.

JV Will you escort me to lat . . . long . . . after refloating?

Leak

JW I have sprung a leak.
JW 1 Leak is dangerous.
JW 2 Leak is causing dangerous heel.
JW 3 Leak is beyond the capacity of my pumps.

I require immediate assistance; I have sprung a leak.   CB 7

JX Leak is gaining rapidly.
JX 1 I cannot stop the leak.

JY Leak can be controlled, if it does not get any worse.
JY 1 I require additional pumping facilities to control the leak.
JY 2 Leak is under control.
JY 3 Leak has been stopped.

JZ Have you sprung a leak?
JZ 1 Can you stop the leak?
JZ 2 Is the leak dangerous?

KA I urgently require a collision mat.
    KA 1 I have placed the collision mat. I can proceed without assistance.
    KA 2 Can you place the collision mat?

KB I have . . . (number) feet or metres of water in the hold.

KC My hold(s) is(are) flooded.
    KC 1 How many compartments are flooded?

KD There are . . . (number) compartments flooded.

KE The watertight bulkheads are standing up well to the pressure of water.
    KE 1 I need timber to support bulkheads.

Towing—tugs

Tug

KF I require a tug (or . . . (number) tugs).

    I require a tug.   Z

KG Do you require a tug(s)?
    KG 1 I do not require tug(s).

KH Tug(s) is(are) coming to you. Expect to arrive at time indicated.
    KH 1 Tug with pilot is coming to you.
    KH 2 You should wait for tugs.

KI There are no tugs available.
    KI 1 Tugs cannot proceed out.

Towing—taking in tow

KJ I am towing a submerged object.
    KJ 1 I am towing a float.
    KJ 2 I am towing a target.

KK Towing is impossible under present weather conditions.
    KK 1 Towing is very difficult.
    KK 2 I cannot connect at present but will attempt when conditions improve.
    KK 3 I cannot connect tonight. I will try in daylight.
    KK 4 Can you assist with your engines?

KL I am obliged to stop towing temporarily.
    KL 1 You should stop towing temporarily.

KM I can take you (or vessel indicated) in tow.
    KM 1 Shall I take you in tow?

KN I cannot take you (or vessel indicated) in tow.
    KN 1 I cannot take you (or vessel indicated) in tow but I will report you and ask for immediate assistance.
    KN 2 I cannot take you (or vessel indicated) in tow but can take off persons.

KO You should endeavour to take vessel (name or identity signal) in tow.
    KO 1 You should report whether you have taken vessel (name or identity signal) in tow.
    KO 2 Can you take me (or vessel indicated) in tow?

KP You should tow me to the nearest port or anchorage (or place indicated).
    KP 1 I will tow you to the nearest port or anchorage (or place indicated).
    KP 2 I must get shelter or anchorage as soon as possible.

KQ Prepare to be taken in tow.
    KQ 1 I am ready to be taken in tow.
    KQ 2 Prepare to tow me (or vessel indicated).
    KQ 3 I am ready to tow you.
    KQ 4 Prepare to resume towing.
    KQ 5 I am ready to resume towing.

KR All is ready for towing.
    KR 1 I am commencing to tow.
    KR 2 You should commence towing.
    KR 3 Is all ready for towing?

Towing line—cable-hawser

KS You should send a line over.
    KS 1 I have taken the line.

KT You should send me a towing hawser.
    KT 1 I am sending towing hawser.

KU I cannot send towing hawser.
    KU 1 I have no, or no other, hawser.
    KU 2 I have no wire hawser.
    KU 3 Have you a hawser?

KV I intend to use my towing hawser/cable.
    KV 1 I intend to use your towing hawser/cable.

KW You should have towing hawser/cable ready.
    KW 1 Towing hawser/cable is ready.
    KW 2 You should have another hawser ready.
    KW 3 You should have spare towing hawser/cable ready.
    KW 4 Spare towing hawser/cable is ready.
    KW 5 You should have wire hawser ready.
    KW 6 Wire hawser is ready.

KX   You should be ready to receive the towing hawser.
    KX 1   I am ready to receive the towing hawser.
    KX 2   You should come closer to receive towing hawser.
    KX 3   I am coming closer to receive towing hawser.
    KX 4   I have received towing hawser.

KY   Length of tow is . . . (number) fathoms.

KZ   You should shorten in the towing hawser (or shorten distance between vessels).
    KZ 1   I am shortening towing hawser (or I am shortening distance between vessels).
    KZ 2   You should haul in the hawser.
    KZ 3   I am hauling in the hawser.
    KZ 4   You should haul in the slack.
    KZ 5   I am hauling in the slack.

LA   Towing hawser/cable has parted.
    LA 1   Towing hawser/cable is in danger of parting.
    LA 2   Towing hawser/cable is damaged.
    LA 3   You should reinforce the hawsers.
    LA 4   I am reinforcing the hawsers.

Make fast—veer

LB   You should make towing hawser fast to your chain cable.
    LB 1   Towing hawser is fast to chain cable.
    LB 2   You should make towing hawser fast to wire.
    LB 3   Towing hawser is fast to wire.
    LB 4   My towing hawser is fast.
    LB 5   Is your towing hawser fast?

LC   You should make fast astern and steer me.

LD   You should veer your hawser/cable (. . . (number) fathoms).

LE   I am about to veer my hawser/cable.
    LE 1   I am veering my hawser/cable.
    LE 2   I have veered my hawser/cable.
    LE 3   I shall veer cable attached to hawser.
    LE 4   How much cable should I veer?

LF   You should stop veering your hawser/cable.
    LF 1   I cannot veer any more hawser/cable.

Cast off

LG   You should prepare to cast off towing hawser(s).
    LG 1   I am preparing to cast off towing hawser(s).
    LG 2   I am ready to cast off towing hawser(s).
    LG 3   You should cast off starboard towing hawser.
    LG 4   I have cast off starboard towing hawser.
    LG 5   You should cast off port towing hawser.
    LG 6   I have cast off port towing hawser.
    LG 7   You should cast off hawser(s).
    LG 8   I must cast off towing hawser(s).

Engine manoeuvres

I am going ahead.	QD
My engines are going ahead.	QD 1
I will keep going ahead.	QD 2
I will go ahead.	QD 3
I will go ahead dead slow.	QD 4
I have headway.	QE
I cannot go ahead.	QF
You should go ahead.	QG
You should go slow ahead.	QG 1
You should go full spead ahead.	QG 2
You should keep going ahead.	QG 3
You should keep your engines going ahead.	QG 4
You should not go ahead any more.	QH
I am going astern.	QI
My engines are going astern.	QI 1
I will keep going astern.	QI 2
I will go astern.	QI 3
I will go astern dead slow.	QI 4
I have sternway.	QJ
I cannot go astern.	QK
You should go astern.	QL
You should go slow astern.	QL 1
You should go full speed astern.	QL 2
You should keep going astern.	QL 3
You should keep your engines going astern.	QL 4
You should not go astern any more.	QM
You should stop your engines immediately.	RL

	You should stop your engines.	RL 1
	My engines are stopped.	RM
	I am stopping my engines.	RM 1
LH	Maximum speed in tow is . . . (number) knots.	
LI	I am increasing speed.	
	LI 1 Increase speed.	
LJ	I am reducing speed.	
	LJ 1 Reduce speed.	

Part III Aids to navigation – navigation – hydrography

Aids to navigation

Buoys—beacons

LK	Buoy (or beacon) has been established in lat . . . long . . .	
LL	Buoy (or beacon) in lat . . . long . . . has been removed.	
	You should steer directly for the buoy (or object indicated).	PL
	You should keep buoy (or object indicated) on your starboard side.	PL 1
	You should keep buoy (or object indicated) on your port side.	PL 2
	You can pass the buoy (or object indicated) on either side.	PL 3
LM	Radio beacon indicated is out of action.	

Lights—lightvessels

LN Light (name follows) has been extinguished.
    LN 1 All lights are out along this coast (or the coast of . . .).

LO I am not in my correct position (to be used by a lightvessel).
    LO 1 Lightvessel (name follows) is out of position.
    LO 2 Lightvessel (name follows) has been removed from her station.

Lightvessel (or lighthouse) indicated requires assistance.	CH 1

Bar

LP There is not less than . . . (number) feet or metres of water over the bar.

LQ There will be . . . (number) feet or metres of water over the bar at time indicated.

LR Bar is not dangerous.
    LR 1 What is the depth of water over the bar?
    LR 2 Can I cross the bar?

LS Bar is dangerous.

Bearings

LT Your bearing from me [or from . . . (name or identity signal)] is . . . (at time indicated).

LU The bearing of . . . (name or identity signal) from . . . (name or identity signal) is . . . (at time indicated).

LV Let me know my bearings from you. I will flash searchlight.
    LV 1 What is my bearing from you [or from . . . (name or identity signal)] ?
    LV 2 What is the bearing of . . . (name or identity signal) from . . . (name or identity signal) ?

Your magnetic bearing from me (or from vessel or position indicated) is . . . (at time indicated).   BZ

What is my magnetic bearing from you (or from vessel or position indicated) ?   CA

LW I receive your transmission on bearing . . .
    LW 1 Can you take bearings from my radio signals?

Your position according to bearings taken by radio direction-finder stations which I control is lat . . . long . . . (at time indicated).   EZ

Will you give me my position according to bearings taken by radio direction-finder stations which you control?   EZ 1

Bearing and distance by radar of vessel (or object) indicated is bearing . . . , distance . . . miles.   OM

What is the bearing and distance by radar of vessel (or object) indicated?   OM 1

Canal—channel—fairway

Canal

LX The canal is clear.
    LX 1 The canal will be clear at time indicated.
    LX 2 You can enter the canal at time indicated.
    LX 3 Is the canal clear?
    LX 4 When can I enter the canal?

LY The canal is not clear.

LZ The channel/fairway is navigable.
    LZ 1 I intend to pass through the channel/fairway.
    LZ 2 Is the channel/fairway navigable?
    LZ 3 What is the state of the channel/fairway?
    LZ 4 What is the least depth of water in the channel/fairway?

MA The least depth of water in the channel/fairway is . . . (number feet or metres).

MB You should keep in the centre of the channel/fairway.
    MB 1 You should keep on the starboard side of the channel/fairway.
    MB 2 You should keep on the port side of the channel/fairway.
    MB 3 You should leave the channel/fairway free.

You appear not to be complying with the traffic separation scheme.   YG

MC	There is an uncharted obstruction in the channel/fairway. You should proceed with caution. MC 1 The channel/fairway is not navigable.		
MC2	The (—) lane of the traffic separation scheme is not navigable. (The direction of the traffic flow is to be indicated).		

Course

MD	My course is... MD 1 What is your course?	
	My present position, course and speed are lat... long...,..., knots...	EV
	What are your present position, course and speed?	EV 1
	Will vessels in my immediate vicinity (or in the vicinity of lat... long...) please indicate position, course and speed.	FB
	Vessel coming to your rescue (or to the rescue of vessel or aircraft indicated) is steering course..., speed...knots.	GR
	You should indicate course and speed of vessel coming to my rescue (or to the rescue of vessel or aircraft indicated).	GR 1
ME	The course to place (name follows) is... ME 1 What is the course to place (name follows)?	
	The magnetic course for the helicopter to regain its base is...	BX
MF	Course to reach me is... MF 1 What is the course to reach you?	
	The magnetic course for you to steer towards me (or vessel or position indicated) is...(at time indicated).	BW
	Will you indicate the magnetic course for me to steer towards you (or vessel or position indicated)?	BY
MG	You should steer course... MG 1 What course should I steer?	
	You should maintain your present course.	PI
	I am maintaining my present course.	PI 1
	I cannot maintain my present course.	PJ
MH	You should alter course to...(at time indicated).	
MI	I am altering course to...	
	I am altering my course to starboard.	E
	I am altering my course to port.	I

	You should alter your course, if possible, appreciably to starboard to facilitate location by radar.	OJ 2
	You should alter your course, if possible, appreciably to port to facilitate location by radar.	OJ 3

Dangers to navigation—warnings

Derelict—wreck—shoal

MJ Derelict dangerous to navigation reported in lat . . . long . . . (or Complements Table III).

MK I have seen derelict (in lat . . . long . . . at time indicated).
MK 1 Have you seen derelict (or wreckage)?

ML Derelict is drifting near lat . . . long . . . (or bearing . . . from place indicated, distance . . .).
ML 1 Hull of derelict is awash.
ML 2 Hull of derelict is well out of the water.

MM There is a wreck in lat . . . long . . .
MM 1 Wreck is buoyed.
MM 2 Wreck is awash.

MN Wreck (in lat . . . long . . .) is not buoyed.

MO I have struck a shoal or submerged object (lat . . . long . . .).

MP I am in shallow water. Please direct me how to navigate.

Radiation danger

MQ There is risk of contamination due to excessive release of radioactive material in this area (or in area around lat . . . long . . .). Keep radio watch. Relay the message to vessels in your vicinity.
MQ 1 The radioactive material is airborne.
MQ 2 The radioactive material is waterborne.

MR There is no, or no more, risk of contamination due to excessive release of radioactive material in this area (or in area around lat . . . long . . .).
MR 1 Is there risk of contamination due to excessive release of radioactive material in this area (or in area around lat . . . long . . .)?

MS My vessel is a dangerous source of radiation.
MS 1 My vessel is a dangerous source of radiation; you may approach from my starboard side.
MS 2 My vessel is a dangerous source of radiation; you may approach from my port side.
MS 3 My vessel is a dangerous source of radiation; you may approach from forward.
MS 4 My vessel is a dangerous source of radiation; you may approach from aft.

MT My vessel is a dangerous source of radiation. You may approach from . . . (Complements table III).

MU My vessel is a dangerous source of radiation. Do not approach within . . . (number) cables.

I am abandoning my vessel which has suffered a nuclear accident and is a possible source of radiation danger.   AD

I have had a serious nuclear accident and you should approach with caution.   AJ

I have had nuclear accident on board.   AK

A vessel which has suffered a nuclear accident is in distress in lat . . . long . . .   EC

MV My vessel is releasing radioactive material and presents a hazard.

MW My vessel is releasing radioactive material and presents a hazard. Do not approach within . . . (number) cables.

MX The radioactive material is airborne. Do not approach from leeward.

Warnings

MY It is dangerous to stop.
    MY 1 It is dangerous to remain in present position.
    MY 2 It is dangerous to proceed on present course.
    MY 3 It is dangerous to proceed until weather permits.
    MY 4 It is dangerous to alter course to starboard.
    MY 5 It is dangerous to alter course to port.
    MY 6 It is dangerous to approach close to my vessel.
    MY 7 It is dangerous to let go an anchor or use bottom trawl.
    MY 8 It is dangerous to jettison inflammable oil.

It is not safe to fire a rocket.   GU

MZ Navigation is dangerous in the area around lat . . . long . . .

    MZ 1   I consider you are carrying out a dangerous navigational practice and I intend to report you.

Navigation is dangerous in the area around lat . . . long . . . owing to iceberg(s).   VZ

Navigation is dangerous in the area around lat . . . long . . . owing to floating ice.   VZ 1

Navigation is dangerous in the area around lat . . . long . . . owing to pack ice.   VZ 2

NA Navigation is closed.
    NA 1 Navigation is possible only with tug assistance.
    NA 2 Navigation is possible only with pilot assistance.

NA 3 Navigation is prohibited within 500 m of this platform.
NA 4 Navigation is prohibited within 500 m of the platform bearing (—) from me.
NA 5 You have been detected navigating within a 500 m Safety Zone (about the platform bearing (—) from me) and will be reported.
NA 6 Anchors with buoys extend up to one mile from this vessel/rig— you should keep clear.

You should navigate with caution. Small fishing boats are within ... (number) miles of me.	TH
You should navigate with caution. You are drifting towards my set of nets.	TI
You should navigate with caution. There are nets with a buoy in this area.	TJ

NB There is fishing gear in the direction you are heading (or in direction indicated—Complements table III).

NC I am in distress and require immediate assistance.

ND Tsunami (phenomenal wave) is expected. You should take appropriate precautions.

Tropical storm (cyclone, hurricane, typhoon) is approaching. You should take appropriate precautions.	VL

NE You should proceed with great caution.
  NE 1 You should proceed with great caution; the coast is dangerous.
  NE 2 You should proceed with great caution; submarines are exercising in this area.
  NE 3 You should proceed with great caution; there is a boom across.
  NE 4 You should proceed with great caution; keep clear of firing range.
  NE 5 You should proceed with great caution; hostile vessel sighted (in lat ... long ...).
  NE 6 You should proceed with great caution; hostile submarine sighted (in lat ... long ...).
  NE 7 You should proceed with great caution; hostile aircraft sighted (in lat ... long ...).

There is an uncharted obstruction in the channel/fairway. You should proceed with caution.	MC
You should change your anchorage/berth. It is not safe.	RE
All vessels should proceed to sea as soon as possible owing to danger in port.	UL
You appear not to be complying with the traffic separation scheme.	YG

NF You are running into danger.
  NF 1 You are running into danger; there is a radiation hazard.

U

NG You are in a dangerous position.
    NG 1 You are in a dangerous position; there is a radiation hazard.
    You appear not to be complying with the traffic separation scheme.   YG

NH You are clear of all danger.
    NH 1 Are you clear of all danger?

NI I have (or vessel indicated has) a list of . . . (number) degrees to starboard.

NJ I have (or vessel indicated has) a list of . . . (number) degrees to port.

Depth—draught

Depth

NK There is not sufficient depth of water.

NL There is sufficient depth of water.
    NL 1 Is there sufficient depth of water?

The least depth of water in the channel/fairway is . . . (number feet or metres).	MA
What is the least depth of water in the channel/fairway?	LZ 4
There is not less than . . . (number feet or metres) water over the bar.	LP
What is the depth of water over the bar?	LR 1
There will be . . . (number feet or metres) of water over the bar at time indicated.	LQ
The depth at high water here (or in place indicated) is . . . (number feet or metres).	QA
The depth at low water here (or in place indicated) is . . . (number feet or metres).	QB
What is the depth at high and low water here (or in place indicated)?	PW 2

NM You should report the depth around your vessel.

NN I am in . . . (number feet or metres) of water.

*NO Negative—"No" or "The significance of the previous group should be read in the negative".

NP The depth of water at the bow is . . . (number feet or metres).

NQ The depth of water at the stern is . . . (number feet or metres).

*Procedural signal.

NR The depth of water along the starboard side is . . . (number feet or metres).

NS The depth of water along the port side is . . . (number feet or metres).

Draught

NT What is your draught?
    NT 1 What is your light draught?
    NT 2 What is your ballast draught?
    NT 3 What is your loaded draught?
    NT 4 What is your summer draught?
    NT 5 What is your winter draught?
    NT 6 What is your maximum draught?
    NT 7 What is your least draught?
    NT 8 What is your draught forward?
    NT 9 What is your draught aft?

NU My draught is . . . (number feet or metres).

NV My light draught is . . . (number feet or metres).

NW My ballast draught is . . . (number feet or metres).

NX My loaded draught is . . . (number feet or metres).

NY My summer draught is . . . (number feet or metres).

NZ My winter draught is . . . (number feet or metres).

OA My maximum draught is . . . (number feet or metres).

OB My least draught is . . . (number feet or metres).

OC My draught forward is . . . (number feet or metres).

OD My draught aft is . . . (number feet or metres).

    My maximum draught when I went aground was (number feet or metres).   JJ

    What was your draught when you went aground?   JI 1

OE Your draught must not exceed . . . (number feet or metres).

OF I could lighten to . . . (number feet or metres) draught.

OG To what draught could you lighten?

Electronic navigation

Radar

OH You should switch on your radar and keep radar watch.
    OH 1 The restrictions on the use of radar are lifted.
    OH 2 Does my radar cause interference?

OI I have no radar.
    OI 1 Are you equipped with radar?
    OI 2 Is your radar in operation?

OJ I have located you on my radar bearing ..., distance ... miles.
    OJ 1 I cannot locate you on my radar.
    OJ 2 You should alter your course, if possible, appreciably to starboard to facilitate location by radar.
    OJ 3 You should alter your course, if possible, appreciably to port to facilitate location by radar.
    OJ 4 Can you locate me by radar?

    My position is ascertained by radar.   EW 4

*OK Acknowledging a correct repetition or "It is correct".

OL Is radar pilotage being effected in this port (or port indicated)?

OM Bearing and distance by radar of vessel, (or object) indicated, is bearing ..., distance ... miles.
    OM 1 What is the bearing and distance by radar of vessel (or object) indicated?

ON I have an echo on my radar on bearing ..., distance ... miles.

Radio direction-finder

OO My radio direction-finder is inoperative.

OP I have requested ... (name or identity signal) to send two dashes of ten seconds each or the carrier of his transmitter followed by his call sign.
    OP 1 Will you request ... (name or identity signal) to send two dashes of ten seconds each or the carrier of his transmitter followed by his call sign?
    OP 2 Will you send two dashes of ten seconds each, or the carrier of your transmitter, followed by your call sign?

    Your position according to bearings taken by radio direction-finder stations which I control is lat ... long ... (at time indicated).   EZ

    Will you give me my position according to bearings taken by by radio direction-finder stations which you control?   EZ 1

*Procedural signal.

OQ I am calibrating radio direction-finder or adjusting compasses.

Decca—Loran—Consol

My position is ascertained by Decca Navigator.   EW 5

My position is ascertained by Loran.   EW 6

My position is ascertained by Consol.   EW 7

Mines—minesweeping

OR I have struck a mine.

    I have a mine in my sweep (or net).   TO

OS There is danger from mines in this area (or area indicated).
    OS 1 You should keep a look-out for mines.
    OS 2 You are out of the dangerous zone.
    OS 3 Am I out of the dangerous zone?
    OS 4 Are you out of the dangerous zone?
    OS 5 Is there any danger from mines in this area (or area indicated)?

OT Mine has been sighted in lat . . . long . . . (or in direction indicated—Complements table III).

OU Mine(s) has(have) been reported in the vicinity (or in approximate position lat . . . long . . .).

OV Mine(s) is(are) believed to be bearing . . . from me, distance . . . miles.

OW There is a minefield ahead of you. You should stop your vessel and wait for instructions.
    OW 1 There is a minefield along the coast. You should not approach too close.

OX The approximate direction of the minefield is bearing . . . from me.

OY Port is mined.
    OY 1 Entrance is mined.
    OY 2 Fairway is mined.
    OY 3 Are there mines in the port, entrance or fairway?

OZ The width of the swept channel is . . . (number feet or metres).

PA I will indicate the swept channel. You should follow in my wake.
    PA 1 You should keep carefully to the swept channel.
    PA 2 The swept channel is marked by buoys.
    PA 3 I do not see the buoys marking the swept channel.
    PA 4 Do you know the swept channel?

*PB You should keep clear of me; I am engaged in minesweeping operations.
    PB 1 You should keep clear of me; I am exploding a floating mine.

PC I have destroyed the drifting mine(s).
    PC 1 I cannot destroy the drifting mine(s).

Navigation lights—searchlight

PD Your navigation light(s) is(are) not visible.
    PD 1 My navigation lights are not functioning.

PE You should extinguish all the lights except the navigation lights.

PG I do not see any light.
    PG 1 You should hoist a light.
    PG 2 I am dazzled by your searchlight. Douse it or lift it.

You should train your searchlight nearly vertical on a cloud, intermittently if possible, and, if my aircraft is seen, deflect the beam up wind and on the water to facilitate my landing.	AX
Shall I train my searchlight nearly vertical on a cloud, intermittently if possible, and, if your aircraft is seen, deflect the beam up wind and on the water to facilitate your landing?	AX 1

Navigating and steering instructions
(see also page 81, Pilot)

PH You should steer as indicated.
    PH 1 You should steer towards me.
    PH 2 I am steering towards you.
    PH 3 You should steer more to starboard.
    PH 4 I am steering more to starboard.
    PH 5 You should steer more to port.
    PH 6 I am steering more to port.

PI You should maintain your present course.
    PI 1 I am maintaining my present course.
    PI 2 Shall I maintain my present course?

PJ I cannot maintain my present course.

You should make fast astern and steer me.	LC

PK I cannot steer without assistance.

---

*The use of this signal does not relieve any vessel from complying with Rule 27(f) of the International Regulations for Preventing Collisions at Sea.

PL   You should steer directly for the buoy (or object indicated).
   PL 1   You should keep buoy (or object indicated) on your starboard side.
   PL 2   You should keep buoy (or object indicated) on your port side.
   PL 3   You can pass the buoy (or object indicated) on either side.

PM   You should follow in my wake (or wake of vessel indicated).
   PM 1   You should go ahead and lead the course.

PN   You should keep to leeward of me (or vessel indicated).
   PN 1   You should keep to windward of me (or vessel indicated).
   PN 2   You should keep on my starboard side (or starboard side of vessel indicated).
   PN 3   You should keep on my port side (or port side of vessel indicated).

PO   You should pass ahead of me (or vessel indicated).
   PO 1   I will pass ahead of you (or vessel indicated).
   PO 2   You should pass astern of me (or vessel indicated).
   PO 3   I will pass astern of you (or vessel indicated).
   PO 4   You should pass to leeward of me (or vessel indicated).
   PO 5   I will pass to leeward of you (or vessel indicated).
   PO 6   You should pass to windward of me (or vessel indicated).
   PO 7   I will pass to windward of you (or vessel indicated).
   PO 8   You should come under my stern.

PP   Keep well clear of me.
   PP 1   Do not overtake me.
   PP 2   Do not pass ahead of me.
   PP 3   Do not pass astern of me.
   PP 4   Do not pass on my starboard side.
   PP 5   Do not pass on my port side.
   PP 6   Do not pass too close to me.
   PP 7   You should give way to me.

PQ   You should keep closer in to the coast.
   PQ 1   You should keep further away from the coast.
   PQ 2   You should follow the coast at a safe distance.
   PQ 3   How far out from the coast?

PR   You should keep closer to me (or vessel indicated).
   PR 1   You should come as near as possible.
   PR 2   You should keep within visual signal distance from me (or vessel indicated).
   PR 3   You should come within hailing distance from me (or vessel indicated).

PS   You should not come any closer.
   PS 1   You should keep away from me (or vessel indicated).

   I am calibrating radio direction-finder or adjusting compasses.   OQ

Tide

PT   What is the state of the tide?
    PT 1   The tide is rising.
    PT 2   The tide is falling.
    PT 3   The tide is slack.

PU   The tide begins to rise at time indicated.
    PU 1   When does the tide begin to rise?

PV   The tide begins to fall at time indicated.
    PV 1   When does the tide begin to fall?

PW   What is the rise and fall of the tide?
    PW 1   What is the set and drift of the tide?
    PW 2   What is the depth at high and low water here (or in place indicated)?

PX   The rise and fall of the tide is . . . (number feet or metres).

PY   The set of the tide is . . . degrees.

PZ   The drift of the tide is . . . knots.

QA   The depth at high water here (or in place indicated) is . . . (number feet or metres).

QB   The depth at low water here (or in place indicated) is . . . (number feet or metres).

The tide was high water when the vessel went aground.	JK
The tide was half water when the vessel went aground.	JK 1
The tide was low water when the vessel went aground.	JK 2
At what state of tide did you go aground?	JI 3

QC   You should wait until high water.
    QC 1   You should wait until low water.

Part IV **Manoeuvres**

Ahead—astern

Ahead—headway

QD I am going ahead.
    QD 1   My engines are going ahead.
    QD 2   I will keep going ahead.
    QD 3   I will go ahead.
    QD 4   I will go ahead dead slow.

QE I have headway.

QF I cannot go ahead.

QG You should go ahead.
    QG 1   You should go slow ahead.
    QG 2   You should go full speed ahead.
    QG 3   You should keep going ahead.
    QG 4   You should keep your engines going ahead.

QH You should not go ahead any more.

Astern—sternway

QI I am going astern.   S
    QI 1   My engines are going astern.
    QI 2   I will keep going astern.
    QI 3   I will go astern.
    QI 4   I will go astern dead slow.

QJ I have sternway.

QK I cannot go astern.

QL You should go astern.
    QL 1   You should go slow astern.
    QL 2   You should go full speed astern.
    QL 3   You should keep going astern.
    QL 4   You should keep your engines going astern.

QM You should not go astern any more.

Alongside

QN You should come alongside my starboard side.
    QN 1   You should come alongside my port side.
    QN 2   You should drop an anchor before coming alongside.

QO You should not come alongside.

QP I will come alongside.
    QP 1   I will try to come alongside.

QQ I require health clearance (see page 94).

QR   I cannot come alongside.
     QR 1 Can I come alongside?

To anchor—anchor(s)—anchorage

To anchor

QS   You should anchor at time indicated.
     QS 1 You should anchor (position to be indicated if necessary).
     QS 2 You should anchor to await tug.
     QS 3 You should anchor with both anchors.
     QS 4 You should anchor as convenient.
     QS 5 Are you going to anchor?

     You should heave to or anchor until pilot arrives.   UB

QT   You should not anchor. You are going to foul my anchor.

QU   Anchoring is prohibited.

QV   I am anchoring in position indicated.
     QV 1 I have anchored with both anchors.

QW   I shall not anchor.
     QW 1 I cannot anchor.

QX   I request permission to anchor.
     QX 1 You have permission to anchor.

QY   I wish to anchor at once.
     QY 1 Where shall I anchor?

Anchor(s)

QZ   You should have your anchors ready for letting go.
     QZ 1 You should let go another anchor.

RA   My anchor is foul.
     RA 1 I have picked up telegraph cable with my anchor.

RB   I am dragging my anchor.   Y
     RB 1 You appear to be dragging your anchor.
     RB 2 Where you have anchored (or intend to anchor) you are
           likely to drag.

RC   I am (or vessel indicated is) breaking adrift.
     RC 1 I have broken adrift.

RD   You should weigh (cut or slip) anchor immediately.
     RD 1 You should weigh anchor at time indicated.
     RD 2 I am unable to weigh my anchor.

Anchorage

RE You should change your anchorage/berth. It is not safe.

RF Will you lead me into a safe anchorage?

    You should tow me to the nearest port or anchorage (or place indicated).   KP

    I will tow you to the nearest port or anchorage (or place indicated).   KP 1

    I must get shelter or anchorage as soon as possible.   KP 2

RG You should send a boat to where I am to anchor or moor.
    RG 1 At what time shall I come into anchorage?

    You should proceed to anchorage in position indicated (lat . . . long . . .).   RW

    You should not proceed out of harbour/anchorage.   RZ 1

RH There is no good holding ground in my area (or around lat . . . long . . .).

RI There is good holding ground in my area (or around lat . . . long . . .).
    RI 1 Is there good holding ground in your area (or around lat . . . long . . .)?

Engines—propeller

Engines

RJ You should keep your engines ready.
    RJ 1 You should have your engines ready as quickly as possible.
    RJ 2 You should report when your engines are ready.
    RJ 3 You should leave when your engines are ready.
    RJ 4 At what time will your engines be ready?

RK My engines will be ready at time indicated.
    RK 1 My engines are ready.

RL You should stop your engines immediately.
    RL 1 You should stop your engines.

RM My engines are stopped.
    RM 1 I am stopping my engines.
    RM 2 I am obliged to stop my engines.

RN My engines are out of action.

    I can only proceed with one engine.   IL 1

Propeller

RO   Propeller shaft is broken.
      RO 1 My propeller is fouled by hawser or rope.
      RO 2 I have lost my propeller.

I require immediate assistance; propeller shaft is broken.   CB 8

Landing—boarding

Landing

*RP   Landing here is highly dangerous.
      *RP 1 Landing here is highly dangerous. A more favourable location for landing is at position indicated.

**RQ   Interrogative or "the significance of the previous group should be read as a question".

*RR   This is the best place to land.
      *RR 1 Lights will be shown or flag waved at the best landing place.

Boat should endeavour to land where flag is waved or light is shown.   DC

Boats are not allowed to land (after time indicated).   DD 1

Boarding

RS   No-one is allowed on board.

You should stop, or heave to, I am going to board you.   SQ 3

Manoeuvre

RT   Stop carrying out your intentions and watch for my signals.   X
      RT 1 What manoeuvres do you intend to carry out?

RU   Keep clear of me; I am manoeuvring with difficulty.   D
      RU 1 I am carrying out manoeuvring trials.

Proceed—under way

Proceed

RV   You should proceed (to place indicated if necessary).
      RV *continued overleaf*

*Reference is made to landing signals prescribed by the International Convention for the Safety of Life at Sea, 1974 (Regulation 16, Chapter V), for the guidance of small boats with crews or persons in distress.
**Procedural signal.

RV 1 You should proceed to destination.
RV 2 You should proceed into port.
RV 3 You should proceed to sea.

RW You should proceed to anchorage in position indicated (lat . . . long . . .).

RX You should proceed at time indicated.

RY You should proceed at slow speed when passing me (or vessels making this signal).

You should proceed to the rescue of vessel (or ditched aircraft) in lat . . . long . . .	GP
You should proceed to lat . . . long . . . to pick up survivors.	HN
You should proceed with great caution.	NE
You should proceed with great caution; the coast is dangerous.	NE 1
You should proceed with great caution; submarines are exercising in this area.	NE 2
You should proceed with great caution; there is a boom across.	NE 3
You should proceed with great caution; keep clear of firing range.	NE 4
You should proceed with great caution; hostile vessel sighted (in lat . . . long . . .).	NE 5
You should proceed with great caution; hostile submarine sighted (in lat . . . long . . .).	NE 6
You should proceed with great caution; hostile aircraft sighted (in lat . . . long . . .).	NE 7

RZ You should not proceed (to place indicated if necessary).
RZ 1 You should not proceed out of harbour/anchorage.

All vessels should proceed to sea as soon as possible owing to danger in port.	UL

SA I can proceed at time indicated.

SB I am proceeding to the position of accident.

I am (or vessel indicated is) proceeding to your assistance.	CP
I am proceeding to the assistance of vessel/aircraft in distress (lat . . . long . . .).	CR
I am proceeding to the position of accident at full speed. Expect to arrive at time indicated.	FE

Are you proceeding to the position of accident? If so, when do you expect to arrive?	FE 1
I cannot proceed to the rescue owing to weather. You should do all you can.	GQ
I will try to proceed by my own means but I request you to keep in contact with me by . . . (Complements table I).	IJ
I can proceed at . . . (number) knots.	IK
I can only proceed at slow speed.	IL
I can only proceed with one engine.	IL 1
I am unable to proceed under my own power.	IL 2
Are you in a condition to proceed?	IL 3
I have placed the collision mat. I can proceed without assistance.	KA 1

Under way

SC   I am under way.
      SC 1 I am ready to get under way.
      SC 2 I shall get under way as soon as the weather permits.

SD   I am not ready to get under way.

SF   Are you (or vessel indicated) under way?
      SF 1 Are you ready to get under way?
      SF 2 At what time will you be under way?

Speed

SG   My present speed is . . . (number) knots.

SJ   My maximum speed is . . . (number) knots.

SL   What is your present speed?
      SL 1 What is your maximum speed?

The speed of my aircraft in relation to the surface of the earth is . . . (knots or kilometres per hr).	BQ
What is the speed of your aircraft in relation to the surface of the earth?	BQ 1
My present position, course and speed are lat . . . long . . ., . . ., knots . . .	EV
What are your present position, course and speed?	EV 1

Will vessels in my immediate vicinity (or in the vicinity of lat . . . long . . .) please indicate position, course and speed.		FB
I can only proceed at slow speed.		IL
Maximum speed in tow is . . . (number) knots.		LH
I am increasing speed.		LI
Increase speed.		LI 1
I am reducing speed.		LJ
Reduce speed.		LJ 1
You should proceed at slow speed when passing me (or vessels making this signal).		RY
Take the way off your vessel.		SP
My vessel is stopped and making no way through the water.		SP 1

SM   I am undergoing speed trials.

Stop—heave to

SN   You should stop immediately. Do not scuttle. Do not lower boats. Do not use the wireless. If you disobey I shall open fire on you.

SO   You should stop your vessel instantly.   L
      SO 1   You should stop. Head off shore.
      SO 2   You should remain where you are.

SP   Take the way off your vessel.
      SP 1   My vessel is stopped and making no way through the water.   M

SQ   You should stop, or heave to.
      SQ 1   You should stop or heave to, otherwise I shall open fire on you.
      SQ 2   You should stop or heave to; I am going to send a boat.
      SQ 3   You should stop, or heave to; I am going to board you.

You should heave to or anchor until pilot arrives.	UB
I am (or vessel indicated is) stopped in thick fog.	XP

Part V **Miscellaneous**

Cargo—ballast

ST What is your cargo?

SU My cargo is agricultural products.
    SU 1 My cargo is coal.
    SU 2 My cargo is dairy products.
    SU 3 My cargo is fruit products.
    SU 4 My cargo is heavy equipment/machinery.
    SU 5 My cargo is livestock.
    SU 6 My cargo is lumber.
    SU 7 My cargo is oil/petroleum products.
    SU 8 I have a general cargo.

SV I am not seaworthy due to shifting of cargo or ballast.

SW I am taking in, or discharging, or carrying dangerous goods.   B

SX You should not discharge oil or oily mixture.

SY The discharge of oil or oily mixture in this area is prohibited within . . . (number) miles from the nearest land.

Crew—persons on board

SZ Total number of persons on board is . . .

TA I have left . . . (number) men on board.

TB . . . (number) persons have died.

TC . . . (number) persons are sick.

I am alighting (in position indicated if necessary) to pick up crew of vessel/aircraft.	AV
I cannot alight but I can lift crew.	AZ
You cannot alight on the deck. Can you lift crew?	BA 1
Boat(s)/raft(s) should approach vessel as near as possible to take off persons.	DA
All persons saved.	GZ
All persons lost.	GZ 1
I (or rescue vessel/aircraft) have rescued . . . (number) injured persons.	HA
Can I transfer rescued persons to you?	HD

*Fishery

- **TD** I am a fishcatch carrier boat.
  - TD 1 I am a mothership for fishing vessel(s).
  - TD 2 Are you a fishing vessel?

- **TE** I am bottom trawling.
  - TE 1 I am trawling with a floating trawl.
  - TE 2 I am long-line fishing.
  - TE 3 I am fishing with towing lines.
  - TE 4 I am engaged on two-boat fishing operation.
  - TE 5 I am drifting on my nets.
  - TE 6 In what type of fishing are you engaged?

- **TF** I am shooting purse seine.
  - TF 1 I am shooting drift nets.
  - TF 2 I am shooting seine net.
  - TF 3 I am shooting trawl.
  - TF 4 I am shooting long lines.

- **TG** I am hauling purse seine.
  - TG 1 I am hauling drift nets.
  - TG 2 I am hauling seine net.
  - TG 3 I am hauling trawl.
  - TG 4 I am hauling long lines.

- **TH** You should navigate with caution. Small fishing boats are within . . . (number) miles of me.

- **TI** You should navigate with caution. You are drifting towards my set of nets.

- **TJ** You should navigate with caution. There are nets with a buoy in this area.

There is fishing gear in the direction you are heading (or in direction indicated—Complements table III).   NB

- **TK** Is there fishing gear set up on my course?

- **TL** My gear is close to the surface in a direction . . . (Complements table III) for a distance of . . . miles.

- **TM** My gear is well below the surface in a direction . . . (Complements table III) for a distance of . . . miles.

- **TN** In what direction, distance and depth does your fishing gear extend?

- **TO** I have a mine in my sweep (or net).

- **TP** Fishing gear has fouled my propeller.

*Displaying any of the signals in this section does not relieve vessels from compliance with the International Regulations for Preventing Collisions at Sea.

TQ   You have caught my fishing gear.
    TQ 1   It is necessary to haul in fishing gear for disentangling.
    TQ 2   I am clearing the fishing gear.
    TQ 3   You should take measures to recover the fishing gear.

TS   You should take the following action with your warps:
    TS 1   Veer the port (stern) warp.
    TS 2   Veer the starboard (fore) warp.
    TS 3   Veer both warps.
    TS 4   Stop veering.
    TS 5   Haul the port (stern) warp.
    TS 6   Haul the starboard (fore) warp.
    TS 7   Haul both warps.
    TS 8   Stop hauling.
    TS 9   You may haul your warps; the trawl is clear.

TU   I have to cut the warps. The trawls are entangled.
    TU 1   Give me your warp. I shall transfer your fishing gear on it.
    TU 2   Your warps are under mine.
    TU 3   Both my warps have parted.
    TU 4   My starboard (fore) warp has parted.
    TU 5   My port (stern) warp has parted.

TV   Fishing in this area is prohibited.
    TV 1   Trawling in this area is dangerous because there is an obstruction.

TW   Attention. You are in the vicinity of prohibited fishery limits.

TX   A fishery protection (or fishery assistance) vessel is in lat . . . long . . .

TY   I request assistance from fishery protection (or fishery assistance) vessel.

TZ   Can you offer assistance? (Complements table II).

Pilot
(*see also page 69, Navigating and steering instructions*)

UA   Pilot will arrive at time indicated.

UB   You should heave to or anchor until pilot arrives.

    I have a pilot on board.   H

UC   Is a pilot available in this place (or place indicated)?

    I require a pilot.   G

UE   Where can I get a pilot (for destination indicated if necessary)?

UF   You should follow pilot boat (or vessel indicated).

UG	You should steer in my wake.	
	You should follow in my wake (or wake of vessel indicated).	PM
	You should go ahead and lead the course.	PM 1
UH	Can you lead me into port?	
UI	Sea is too rough; pilot boat cannot get off to you.	
UJ	Make a starboard lee for the pilot boat. UJ 1 Make a port lee for the pilot boat.	
UK	Pilot boat is most likely on bearing...from you. UK 1 Have you seen the pilot boat?	
	Is radar pilotage being effected in this port (or port indicated)?	QL

Port—harbour

UL	All vessels should proceed to sea as soon as possible owing to danger in port.	
UM	The harbour (or port indicated) is closed to traffic.	
	You should not proceed out of harbour/anchorage.	RZ 1
UN	You may enter the harbour immediately (or at time indicated).	
	UN 1 May I enter harbour?	
	UN 2 May I leave harbour?	
UO	You must not enter harbour.	
UP	Permission to enter harbour is urgently requested. I have an emergency case.	
	You should proceed into port.	RV 2
	Can you lead me into port?	UH
UQ	You should wait outside the harbour (or river mouth). UQ 1 You should wait outside the harbour until daylight.	
UR	My estimated time of arrival (at place indicated) is (time indicated). UR 1 What is your estimated time of arrival (at place indicated)?	

Miscellaneous

US Nothing can be done until time indicated.
    US 1 Nothing can be done until daylight.
    US 2 Nothing can be done until tide has risen.
    US 3 Nothing can be done until visibility improves.
    US 4 Nothing can be done until weather moderates.
    US 5 Nothing can be done until draught is lightened.
    US 6 Nothing can be done until tugs have arrived.

UT Where are you bound for?
    UT 1 Where are you coming from?

UU I am bound for...

UV I am coming from...

UW I wish you a pleasant voyage.
    UW 1 Thank you very much for your co-operation. I wish you a pleasant voyage.

UX No information available.

I am unable to answer your question.   YK

Exercises

UY I am carrying out exercises. Please keep clear of me.

Bunkers

UZ I have bunkers for...(number) hrs.

VB Have you sufficient bunkers to reach port?

VC Where is the nearest place at which fuel oil is available?
    VC 1 Where is the nearest place at which diesel oil is available?
    VC 2 Where is the nearest place at which coal is available?

VD Bunkers are available at place indicated (or lat...long...).

Fumigation

VE I am fumigating my vessel.

No-one is allowed on board.   RS

Identification

What is the name or identity signal of your vessel (or station)?   CS

VF You should hoist your identity signal.

Part VI Meteorology—weather

Clouds

VG The coverage of low clouds is . . . (number of octants or eighths of sky covered).

VH The estimated height of base of low clouds in hundreds of metres is . . .

VI What is the coverage of low clouds in octants (eighths of sky covered)?
    VI 1 What is the estimated height of base of low clouds in hundreds of metres?

Gale—storm—tropical storm

Gale

VJ Gale (wind force Beaufort 8–9) is expected from direction indicated (Complements Table III).

Storm

VK Storm (wind force Beaufort 10 or above) is expected from direction indicated (Complements Table III).

Tropical storm

VL Tropical storm (cyclone, hurricane, typhoon) is approaching. You should take appropriate precautions.

VM Tropical storm is centred at . . . (time indicated) in lat . . . long . . . on course . . . , speed . . . knots.

VN Have you latest information of the tropical storm (near lat . . . long . . . if necessary)?

Very deep depression is approaching from direction indicated (Complements Table III).   WT

There are indications of an intense depression forming in lat . . . long . . .   WU

Ice—icebergs

Ice

VO Have you encountered ice?

VP What is the character of ice, its development and the effects on navigation?

VQ Character of ice:
  VQ 0 No ice.
  VQ 1 New ice (ice crystals, slush or sludge, pancake ice or ice rind).
  VQ 2 Young fast ice (5–15 cm. thick or rotten fast ice).
  VQ 3 Open drift ice (not more than 5/8 of the water surface is covered by ice-floes).
  VQ 4 A compressed accumulation of sludge (a compressed mass of sludge or pancake ice, the ice cannot spread).
  VQ 5 Winter fast ice (more than 15 cm. in thickness).
  VQ 6 Close drift ice (the area is covered by ice-floes to a greater extent than 5/8).
  VQ 7 Very close drift ice on open sea.
  VQ 8 Pressure ice or big, vast, heavy ice-floes.
  VQ 9 Shore lead along the coast.

  No information available.   UX

VR Ice development:
  VR 0 No change.
  VR 1 Ice situation has improved.
  VR 2 Ice situation has deteriorated.
  VR 3 Ice has been broken up.
  VR 4 Ice has opened or drifted away.
  VR 5 New ice has been formed and/or the thickness of the ice has increased.
  VR 6 Ice has been frozen together.
  VR 7 Ice has drifted into the area or has been squeezed together.
  VR 8 Warning of pressure ridges.
  VR 9 Warning of hummocking or ice screwing.

  No information available.   UX

VS Effects of the ice on navigation:
  VS 0 Unobstructed.
  VS 1 Unobstructed for power-driven vessels built of iron or steel, dangerous for wooden vessels without ice protection.
  VS 2 Difficult for low-powered vessels without the assistance of an ice-breaker, dangerous for vessels of weak construction.
  VS 3 Possible without ice-breaker only for highly-powered vessels of strong construction.
  VS 4 Ice-breaker assistance available in case of need.
  VS 5 Proceed in channel without the assistance of ice-breaker.
  VS 6 Possible only with the assistance of an ice-breaker.
  VS 7 Ice-breaker can give assistance only to ships strengthened for navigation in ice.
  VS 8 Temporarily closed for navigation.
  VS 9 Navigation has ceased.

  No information available.   UX

VT Danger of ice accretion on superstructure (for example black frost).
  VT 1 I am experiencing heavy icing on superstructure.

VU I have seen ice-field in lat . . . long . . .

VV Ice patrol ship is not on station.
   VV 1 Ice patrol ship is on station.

Icebergs

VW I have seen icebergs in lat . . . long . . .

VX I have encountered one or more icebergs or growlers (with or without position and time).

VY One or more icebergs or growlers have been reported (with or without position and time).

VZ Navigation is dangerous in the area around lat . . . long . . . owing to iceberg(s).
   VZ 1 Navigation is dangerous in the area around lat . . . long . . . owing to floating ice.
   VZ 2 Navigation is dangerous in the area around lat . . . long . . . owing to pack ice.

*Ice-breaker

†WA Repeat word or group after . . .

†WB Repeat word or group before . . .

WC I am (or vessel indicated is) fast on ice and require(s) ice-breaker assistance.
   WC 1 Ice-breaker is being sent to your assistance.

   I require assistance in the nature of ice-breaker.   CD 9

WD Ice-breaker is not available.
   WD 1 Ice-breaker cannot render assistance at present.

WE Navigation channel is being kept open by ice-breaker.

WF I can give ice-breaker support only up to lat . . . long . . .

WG Open channel or open area is in the direction in which aircraft is flying.

WH I can only assist if you will make all efforts to follow.

WI At what time will you follow at full speed?

WJ The convoy will start at time indicated from here (or from lat . . . long . . .).

WK You (or vessel indicated) will be number . . . in convoy.

*Special single-letter signals for use between ice-breakers and assisted vessels can be found on pages 24–25.
†Procedural signals.

WL Ice-breaker is stopping work during darkness.

*WM Ice-breaker support is now commencing. Use special ice-breaker support signals and keep continuous watch for sound, visual or radiotelephony signals.

WN Ice-breaker is stopping work for . . . (number) hrs or until more favourable conditions arise.

WO Ice-breaker support is finished. Proceed to your destination.

You should go astern.   QL

Atmospheric pressure—temperature

Atmospheric pressure

WP Barometer is steady.
    WP 1 Barometer is falling rapidly.
    WP 2 Barometer is rising rapidly.

WQ The barometer has fallen . . . (number) millibars during the past three hours.

WR The barometer has risen . . . (number) millibars during the past three hours.

WS Corrected atmospheric pressure at sea level is . . . (number) millibars.
    WS 1 State corrected atmospheric pressure at sea level in millibars.

WT Very deep depression is approaching from direction indicated (Complements Table III).

WU There are indications of an intense depression forming in lat . . . long . . .

Temperature

WV The air temperature is sub-zero (centigrade).
    WV 1 The air temperature is expected to be sub-zero (centigrade).

Sea—swell

Sea

WW What are the sea conditions in your area (or around lat . . . long . . .)?

---

*Special single-letter signals for use between ice-breakers and assisted vessels can be found on pages 24–25.

WX The true direction of the sea in tens of degrees is ... (number following indicates tens of degrees).

WY The state of the sea is ... (Complements 0–9 corresponding to the following table):

	Height	
	in metres	in feet
0 Calm (glassy)	0	0
1 Calm (rippled)	0 – 0.1	0 – $\frac{1}{3}$
2 Smooth (wavelets)	0.1 – 0.5	$\frac{1}{3}$ – $1\frac{2}{3}$
3 Slight	0.5 – 1.25	$1\frac{2}{3}$ – 4
4 Moderate	1.25 – 2.5	4 – 8
5 Rough	2.5 – 4	8 – 13
6 Very rough	4 – 6	13 – 20
7 High	6 – 9	20 – 30
8 Very high	9 – 14	30 – 45
9 Phenomenal	over 14	over 45

WZ What are the forecast sea conditions in my area (or area around lat ... long ...)?

XA The true direction of the sea in tens of degrees is expected to be ... (number following indicates tens of degrees).

XB The state of the sea is expected to be ... (Complements 0–9 as in the table above).

Swell

XC What are the swell conditions in your area (or area around lat ... long ...)?

XD The true direction of the swell in tens of degrees is ... (number following indicates tens of degrees).

XE The state of the swell is ... (Complements 0–9 corresponding to the following table):

0 No swell
1 Short or middle } weak—approximate height <2 m. (6 ft.)
2 Long

3 Short
4 Middle } moderate—approximate height 2–4 m. (6–12 ft.)
5 Long

6 Short
7 Middle } high—approximate height >4 m. (12 ft.)
8 Long

9 Confused

XF What are the forecast swell conditions in my area (or area around lat ... long ...)?

XG—XU 89

XG The true direction of the swell in tens of degrees is expected to be ... (number following indicates tens of degrees).

XH The state of the swell is expected to be ... (Complements 0–9 as in the table at XE above).

Tsunami (phenomenal wave) is expected. You should take appropriate precautions. ND

Visibility—fog

XI Indicate visibility.

XJ Visibility is ... (number) tenths of nautical miles.

XK Visibility is variable between ... and ... (maximum and minimum in tenths of nautical miles).

XL Visibility is decreasing.
    XL 1 Visibility is increasing.
    XL 2 Visibility is variable.

XM What is the forecast visibility in my area (or area around lat ... long ...)?

XN Visibility is expected to be ... (number) tenths of nautical miles.

XO Visibility is expected to decrease.
    XO 1 Visibility is expected to increase.
    XO 2 Visibility is expected to be variable.

XP I am (or vessel indicated is) stopped in thick fog.
    XP 1 I am entering zone of restricted visibility.

Weather—weather forecast

XQ What weather are you experiencing?

XR Weather is good.
    XR 1 Weather is bad.
    XR 2 Weather is moderating.
    XR 3 Weather is deteriorating.

XS Weather report is not available.

XT Weather expected is bad.
    XT 1 Weather expected is good.
    XT 2 No change is expected in the weather.
    XT 3 What weather is expected?

XU You should wait until the weather moderates.
    XU ♦ I will wait until the weather moderates.

XV   Please give weather forecast for my area (or area around lat . . . long . . .) in MAFOR Code.

Wind

XW   What is the true direction and force of wind in your area (or area around lat . . . long . . .)?

XX   True direction of wind is . . . (Complements table III).

XY   Wind force is Beaufort Scale . . . (numerals 0—12).

XZ   What is the wind doing?
    XZ 1   The wind is backing.
    XZ 2   The wind is veering.
    XZ 3   The wind is increasing.
    XZ 4   The wind is squally.
    XZ 5   The wind is steady in force.
    XZ 6   The wind is moderating.

YA   What wind direction and force is expected in my area (or area around lat . . . long . . .)?

YB   True direction of wind is expected to be . . . (Complements table III).

YC   Wind force expected is Beaufort Scale . . . (numerals 0—12).

YD   What is the wind expected to do?
    YD 1   The wind is expected to back.
    YD 2   The wind is expected to veer.
    YD 3   The wind is expected to increase.
    YD 4   The wind is expected to become squally.
    YD 5   The wind is expected to remain steady in force.
    YD 6   The wind is expected to moderate.

Part VII **Routeing of Ships**

YG You appear not to be complying with the traffic separation scheme.

Part VIII Communications

Acknowledge—answer

YH I have received the following from . . . (name or identity signal of vessel or station).

YI I have received the safety signal sent by . . . (name or identity signal).

YJ Have you received the safety signal sent by . . . (name or identity signal)?

YK I am unable to answer your question.

Received, or I have received your last signal. R (procedure signal)

Calling

YL I will call you again at . . . hrs (on . . . kHz or MHz).

YM Who is calling me?

Cancel

YN Cancel my last signal/message.

My last signal was incorrect. I will repeat it correctly. ZP

Communicate

I wish to communicate with you by . . . (Complements table I). K (with one numeral)

I wish to communicate with you. K

YO I am going to communicate by . . . (Complements table I).

YP I wish to communicate with vessel or coast station (identity signal) by . . . (Complements table I).

YQ I wish to communicate by . . . (Complements table I) with vessel bearing . . . from me.

YR Can you communicate by . . . (Complements table I)?

YS I am unable to communicate by . . . (Complements table I).

YT I cannot read your . . . (Complements table I).

*YU I am going to communicate with your station by means of the International Code of Signals.

*YV The groups which follow are from the International Code of Signals.
   YV 1 The groups which follow are from the local code.

YW I wish to communicate by radiotelegraphy on frequency indicated.

YX I wish to communicate by radiotelephony on frequency indicated.

YY I wish to communicate by VHF radiotelephony on channel indicated.

YZ The words which follow are in plain language.

ZA I wish to communicate with you in . . . (language indicated by following complements).
   0 Dutch   5 Italian
   1 English   6 Japanese
   2 French   7 Norwegian
   3 German   8 Russian
   4 Greek   9 Spanish

ZB I can communicate with you in language indicated (complements as above).

ZC Can you communicate with me in language indicated (complements as above)?

ZD Please communicate the following to all shipping in the vicinity.
   ZD 1 Please report me to Coast Guard New York.
   ZD 2 Please report me to Lloyd's London.
   ZD 3 Please report me to Minmorflot Moscow.
   ZD 4 Please report me to M S A Tokyo.

ZE You should come within visual signal distance.

You should keep within visual signal distance from me (or vessel indicated).	PR 2
I have established communications with the aircraft in distress on 2182 kHz.	BC
Can you communicate with the aircraft?	BC 1
I have established communications with the aircraft in distress on . . . KHz	BD
I have established communications with the aircraft in distress on . . . MHz.	BE

*The abbreviation INTERCO may also be used to mean: "International Code group(s) follow(s)".

Exercise

ZF I wish to exercise signals with you by . . . (Complements table I).

ZG It is not convenient to exercise signals.

ZH Exercise has been completed.

Reception—transmission

ZI I can receive but not transmit by . . . (Complements table I).

ZJ I can transmit but not receive by . . . (Complements table I).

ZK I cannot distinguish your signal. Please repeat it by . . . (Complements table I).

ZL Your signal has been received but not understood.

    I cannot read your . . . (Complements table I).   YT

ZM You should send (or speak) more slowly.
    ZM 1 Shall I send (or speak) more slowly?

ZN You should send each word or group twice.

ZO You should stop sending.
    ZO 1 Shall I stop sending?

Repeat

ZP My last signal was incorrect. I will repeat it correctly.

ZQ Your signal appears incorrectly coded. You should check and repeat the whole.

ZR Repeat the signal now being made to me by vessel (or coast station) . . . (name or identity signal).

Part IX International Health Regulations

Pratique messages

ZS	My vessel is "healthy" and I request free pratique.	Q
	*I require health clearance.	QQ
ZT	My Maritime Declaration of Health has negative answers to the six health questions.	
ZU	My Maritime Declaration of Health has a positive answer to question(s) ... (indicated by appropriate number(s)).	
ZV	I believe I have been in an infected area during the last thirty days.	
ZW	I require Port Medical Officer. ZW 1 Port Medical Officer will be available at (time indicated).	
ZX	You should make the appropriate pratique signal.	
ZY	You have pratique.	
ZZ	You should proceed to anchorage for health clearance (at place indicated). ZZ 1 Where is the anchorage for health clearance?	
	I have a doctor on board.	AL
	Have you a doctor?	AM

*By night, a red light over a white light may be shown where it can best be seen by vessels requiring health clearance. These lights should **only** be about two metres (6 feet) apart, should be exhibited within the precincts of a port, and should be visible all round the horizon **as nearly as possible.**

Tables of complements

Table I

1. Semaphore
2. Morse signalling by hand-flags or arms
3. Loud hailer (megaphone)
4. Morse signalling lamp
5. Sound signals
6. International Code flags
7. Radiotelegraphy 500 kHz
8. Radiotelephony 2182 kHz
9. VHF Radiotelephony—channel 16

Table II

0. Water
1. Provisions
2. Fuel
3. Pumping equipment
4. Fire-fighting appliances
5. Medical assistance
6. Towing
7. Survival craft
8. Vessel to stand by
9. Ice-breaker

Table III

0. Direction unknown (or calm)
1. North-east
2. East
3. South-east
4. South
5. South-west
6. West
7. North-west
8. North
9. All directions (or confused or variable)

Medical Section

Instructions

A. General

1. Medical advice should be sought and given in plain language whenever it is possible but, if language difficulties are encountered, this Code should be used.

2. Even when plain language is used, the text of the Code and the instructions should be followed as far as possible.

3. Reference is made to the procedure signals "C", "N" or "NO" and "RQ" which, when used after the main signal, change its meaning into affirmative, negative and interrogative respectively (see paragraph 3 (j), Chapter VI, page 10). Example:
"MFE N" = "Bleeding is not severe"
"MFE RQ" = "Is bleeding severe?".

B. Instructions to masters

Standard method of case description

1. The master should make a careful examination of the patient and should try to collect, as far as possible, information covering the following subjects:
   (a) Description of the patient (page 102);
   (b) Previous health (page 103);
   (c) Localization of symptoms, diseases or injuries (page 103);
   (d) General symptoms (page 103)
   (e) Particular symptoms (page 107);
   *(f) Diagnosis (page 118).

2. Such information should be coded by choosing the appropriate groups from the corresponding chapters of this section. It would help the recipients of the signal if the information is transmitted in the order stated in paragraph 1.

3. Page 102 contains signals which can be used independently, i.e. with or without the description of the case.

4. After a reply from the doctor has been received and the instructions therein followed, the master can give a progress report by using signals from page 117.

C. Instructions to doctors

1. Additional information can be requested by using page 118.
Example:
"MQB" = "I cannot understand your signal, please use standard method of case description".

2. For diagnosis*, Page 118 should be used. Example:
"MQE 26"   "My probable diagnosis is cystitis".

---

*"Diagnosis" Page 118, can be used by both the master ("request for medical assistance") and the doctor ("medical advice").

3. Prescribing should be limited to the "List of Medicaments" which is set out in Table M III of the Code.

4. For special treatment, signals from page 118 should be used. Example:
"MRP 4" ="Apply ice-cold compress and renew every 4 hours".

5. When prescribing a medicament (page 120) three signals should be used as follows:
(a) the first ("Prescripting", page 120 and Table M III) to signify the medicament itself. Example:
"MTD 32"   "You should give aspirin tablets".
(b) the second ("Methods of administration and dose", page 120) to signify the method of administration and dose. Example:
"MTI 2"   "You should give by mouth 2 tablets/capsules".
(c) the third ("Frequency of dose", page 120) to signify the frequency of the dose. Example:
"MTQ 8"   "You should repeat every 8 hours".

6. The frequency of external applications is set out in "Frequency of external application", page 121. Example:
"MTU 4"   "You should apply every 4 hours".

7. Advice concerning diet can be given by using signals from "Diet", page 121. Example:
"MUC"   "Give water only in small quantities".

D. Examples

As an example, two cases of request for assistance and the corresponding replies are drafted below:

Case One **Request for medical assistance**

"I have a male age (44) years. Patient has been ill for (2) days. Patient has suffered from (bronchitis acute). Onset was sudden. Patient is delirious. Patient has fits of shivering. Temperature taken in mouth is (40). Pulse rate per minute is (110). The rate of breathing per minute is (30). Patient is in pain (chest). Part of the body affected is right (chest). Pain is increased on breathing. Patient has severe cough. Patient has bloodstained sputum. Patient has been given (penicillin injection) without effect. Patient has received treatment by medicaments in last (18) hours. My probable diagnosis is (pneumonia)."

Medical advice

"Your diagnosis is probably right. You should continue giving (penicillin injection). You should repeat every (12) hours. Put patient to bed lying down at absolute rest. Keep patient warm. Give fluid diet, milk, fruit juice, tea, mineral water. Give water very freely. Refer back to me in (24) hours or before if patient worsens."

Case Two   Request for medical assistance

"I have a male aged (31) years. Patient has been ill for (3) hours. Patient has had no serious previous illness. Pulse rate per minute is (95). Pulse is weak. Patient is sweating. Patient is in pain in lumbar (kidney) region. The part affected is left lumbar (kidney) region. Pain is severe. Pain is increased by hand pressure. Bowels are regular.

Request for additional information

"I cannot make a diagnosis. Please answer the following question(s). Temperature taken in mouth is (number). Pain radiates to groin and testicle. Patient has pain on passing water. Urinary functions normal. Vomiting is present."

Additional information

"Temperature taken in mouth is (37). Pain radiates to groin and testicle. Patient has pain on passing water. Patient is passing small quantities of urine frequently. Vomiting is absent. Patient has nausea."

Medical advice

"My probable diagnosis is kidney stone (renal colic). You should give morphine injection. You should give by subcutaneous injection (15) milligrams. Give water freely. Apply hot water bottle to lumber (kidney) region. Patient should be seen by doctor when next in port."

Part I Request for medical assistance

Request—general information

MAA I request urgent medical advice.

MAB I request you to make rendezvous in position indicated.

MAC I request you to arrange hospital admission.

MAD I am . . . (indicate number) hours from nearest port.

MAE I am converging on nearest port.

MAF I am moving away from nearest port.

I require medical assistance.	W
I have a doctor on board.	AL
Have you a doctor?	AM
I need a doctor.	AN
I need a doctor; I have severe burns.	AN 1
I need a doctor; I have radiation casualties.	AN 2
I require a helicopter urgently with a doctor.	BR 2
I require a helicopter urgently to pick up injured/sick person.	BR 3
Helicopter is coming to you now (or at time indicated) with a doctor.	BT 2
Helicopter is coming to you now (or at time indicated) to pick up injured/sick person.	BT 3
I have injured/sick person (or number of persons indicated) to be taken off urgently.	AQ
You should send a helicopter/boat with a stretcher.	BS
A helicopter/boat is coming to take injured/sick.	BU
You should send injured/sick persons to me.	AT

Description of patient

MAJ I have a male aged . . . (number) years.

MAK I have a female aged . . . (number) years.

MAL I have a female . . . (number) months pregnant.

MAM Patient has been ill for . . . (number) days.

MAN   Patient has been ill for . . . (number) hours.

MAO   General condition of the patient is good.

MAP   General condition of the patient is serious.

MAQ   General condition of the patient is unchanged.

MAR   General condition of the patient has worsened.

MAS   Patient has been given . . . (Table M III) with effect.

MAT   Patient has been given . . . (Table M III) without effect.

MAU   Patient has received treatment by medicaments in last . . . (indicate number) hours.

Previous health

MBA   Patient has suffered from . . . (Table M II).

MBB   Patient has had previous operation . . . (Table M II).

MBC   Patient has had no serious previous illness.

MBD   Patient has had no relevant previous injury.

Localization of symptoms, diseases or injuries

MBE   The whole body is affected.

MBF   The part of the body affected is . . . (Table M I).

*MBG   The part of the body affected is right . . . (Table M I).

*MBH   The part of the body affected is left . . . (Table M I).

General symptoms

MBP   Onset was sudden.

MBQ   Onset was gradual.

Temperature

MBR   Temperature taken in mouth is . . . (number).

MBS   Temperature taken in rectum is . . . (number).

*To be used when right and left side of the body or limb need to be differentiated.

MBT	Temperature in the morning is . . . (number).
MBU	Temperature in the evening is . . . (number).
MBV	Temperature is rising.
MBW	Temperature is falling.

Pulse

MBX	The pulse rate per minute is . . . (number).
MBY	The pulse rate is irregular.
MBZ	The pulse rate is rising.
MCA	The pulse rate is falling.
MCB	The pulse is weak.
MCC	The pulse is too weak to count.
MCD	The pulse is too rapid to count.

Breathing

MCE	The rate of breathing per minute is . . . (number) (in and out being counted as one breath).
MCF	The breathing is weak.
MCG	The breathing is wheezing.
MCH	The breathing is regular.
MCI	The breathing is irregular.
MCJ	The breathing is strenuous (noisy).

Sweating

MCL	Patient is sweating.
MCM	Patient has fits of shivering (chills).
MCN	Patient has night sweats.
MCO	Patient's skin is hot and dry.
MCP	Patient is cold and clammy.

Mental state and consciousness

MCR Patient is conscious.

MCT Patient is semi-conscious but can be roused.

MCU Patient is unconscious.

MCV Patient found unconscious.

MCW Patient appears to be in a state of shock.

MCX Patient is delirious.

MCY Patient has mental symptoms.

MCZ Patient is paralyzed . . . (Table M I).

MDC Patient is restless.

MDD Patient is unable to sleep.

Pain

MDF Patient is in pain . . . (Table M I).

MDG Pain is a dull ache.

MDJ Pain is slight.

MDL Pain is severe.

MDM Pain is intermittent.

MDN Pain is continuous.

MDO Pain is increased by hand pressure.

MDP Pain radiates to . . . (Table M I).

MDQ Pain is increased on breathing.

MDR Pain is increased by action of bowels.

MDT Pain is increased on passing water.

MDU Pain occurs after taking food.

MDV Pain is relieved by taking food.

MDW Pain has no relation to taking food.

MDX Pain is relieved by heat.

MDY Pain has ceased.

Cough

MED   Cough is present.

MEF   Cough is absent.

Bowels

MEG   Bowels are regular.

MEJ   Patient is constipated and bowels last opened . . . (indicate number of days).

MEL   Patient has diarrhoea . . . (indicate number of times daily).

Vomiting

MEM   Vomiting is present.

MEN   Vomiting is absent.

MEO   Patient has nausea.

Urine

MEP   Urinary functions normal.

MEQ   Urinary functions abnormal.

Bleeding

MER   Bleeding is present . . . (Table M I).

MET   Bleeding is absent.

Rash

MEU   A rash is present . . . (Table M I).

MEV   A rash is absent.

Swelling

MEW   Patient has a swelling . . . (Table M I).

MEX   Swelling is hard.

MEY   Swelling is soft.

MEZ   Swelling is hot and red.

MFA Swelling is painful on hand pressure.

MFB Swelling is discharging.

MFC Patient has an abscess . . . (Table M I).

MFD Patient has a carbuncle . . . (Table M I).

Particular symptoms

Accidents, injuries, fractures, suicide and poisons

	Bleeding is present . . . (Table M I).	MER

MFE Bleeding is severe.

MFF Bleeding is slight.

MFG Bleeding has been stopped by pad(s) and bandaging.

MFH Bleeding has been stopped by tourniquet.

MFI Bleeding has stopped.

MFJ Bleeding cannot be stopped.

MFK Patient has a superficial wound . . . (Table M I).

MFL Patient has a deep wound . . . (Table M I).

MFM Patient has penetrating wound . . . (Table M I).

MFN Patient has a clean-cut wound . . . (Table M I).

MFO Patient has a wound with ragged edges . . . (Table M I).

MFP Patient has a wound discharging . . . (Table M I).

MFQ Patient has contusion (bruising) . . . (Table M I).

MFR Wound is due to blow.

MFS Wound is due to crushing.

MFT Wound is due to explosion.

MFU Wound is due to fall.

MFV Wound is due to gun-shot.

MFW Patient has a foreign body in wound.

MFX Patient is suffering from concussion.

MFY Patient cannot move the arm . . . (Table M I).

MFZ Patient cannot move the leg . . . (Table M I).

MGA Patient has dislocation . . . (Table M I).

MGB Patient has simple fracture . . . (Table M I).

MGC Patient has compound fracture . . . (Table M I).

MGD Patient has comminuted fracture . . . (Table M I).

MGE Patient has attempted suicide.

MGF Patient has cut throat.

MGG Patient has superficial burn . . . (Table M I).

MGH Patient has severe burn . . . (Table M I).

MGI Patient is suffering from non-corrosive poisoning (no staining and burning of mouth and lips).

MGJ Patient has swallowed corrosive (staining and burning of mouth and lips).

MGK Patient has swallowed unknown poison.

MGL Patient has swallowed a foreign body.

MGM Emetic has been given with good results.

MGN Emetic has been given without good results.

MGO No emetic has been given.

MGP Patient has had corrosive thrown on him . . . (Table M I).

MGQ Patient has inhaled poisonous gases, vapours, dust.

MGR Patient is suffering from animal bite . . . (Table M I).

MGS Patient is suffering from snake bite . . . (Table M I).

MGT Patient is suffering from gangrene . . . (Table M I).

Diseases of nose and throat

MGU Patient has nasal discharge.

MGV Patient has foreign body in nose.

MHA Lips are swollen.

MHB Tongue is dry.

MHC Tongue is coated.

MHD	Tongue is glazed and red.	
MHF	Tongue is swollen.	
MHG	Patient has ulcer on tongue.	
MHJ	Patient has ulcer in mouth.	
MHK	Gums are sore and bleeding.	
MHL	Throat is sore and red.	
MHM	Throat has pinpoint white spots on tonsils.	
MHN	Throat has grey white patches on tonsils.	
MHO	Throat hurts and is swollen on one side.	
MHP	Throat hurts and is swollen on both sides.	
MHQ	Swallowing is painful.	
MHR	Patient cannot swallow.	
MHT	Patient has hoarseness of voice.	
	Patient has swallowed a foreign body.	MGL
MHV	Patient has severe toothache.	

Diseases of respiratory system

MHY	Patient has pain in chest on breathing ... (Table M I).	
	Breathing is wheezing.	MCG
MHZ	Breathing is deep.	
MIA	Patient has severe shortness of breath.	
MIB	Patient has asthmatical attack.	
	Cough is absent.	MEF
MIC	Patient has severe cough.	
MID	Cough is long-standing.	
MIF	Patient is coughing up blood.	
MIG	Patient has no sputum.	
MIJ	Patient has abundant sputum.	
MIK	Sputum is offensive.	

MIL	Patient has blood-stained sputum.	
MIM	Patient has blueness of face.	

Diseases of the digestive system

MIN	Patient has tarry stool.	
MIO	Patient has clay-coloured stool.	
	Patient has diarrhoea . . . (indicate number of times daily).	MEL
MIP	Patient has diarrhoea with frequent stools like rice water.	
MIQ	Patient is passing blood with stools.	
MIR	Patient is passing mucus with stools.	
	Patient has nausea	MEO
MIT	Patient has persistent hiccough	
MIU	Patient has cramp pains and vomiting.	
	Vomiting is present	MEM
	Vomiting is absent	MEN
MIV	Vomiting has stopped.	
MIW	Vomiting is persistent.	
MIX	Vomit is streaked with blood.	
MIY	Patient vomiting much blood.	
MIZ	Vomit is dark (like coffee grounds).	
MJA	Patient vomits any food and liquid given.	
MJB	Amount of vomit is . . . (indicate in decilitres: 1 decilitre equals one-sixth of a pint).	
MJC	Frequency of vomiting is . . . (indicate number) daily.	
MJD	Patient has flatulence.	
MJE	Wind has not been passed per anus for . . . (indicate number of hours).	
MJF	Wind is being passed per anus.	
MJG	Abdomen is distended.	
MJH	Abdominal wall is soft (normal).	

MJI	Abdominal wall is hard and rigid.	
MJJ	Abdominal wall is tender . . . (Table M I).	
	Patient is in pain . . . (Table M I).	MDF
	Patient has a swelling . . . (Table M I).	MEW
MJK	Hernia is present.	
MJM	Hernia cannot be replaced.	
MJN	Hernia is painful and tender.	
MJO	Patient has bleeding haemorrhoids.	
MJP	Haemorrhoids cannot be reduced (put back in place).	

Diseases of the genito-urinary system

	Patient is in pain . . . (Table M I).	MDF
MJS	Patient has pain on passing water.	
MJT	Patient has pain in penis at end of passing water.	
MJU	Patient has pain spreading from abdomen to penis, testicles or thigh.	
MJV	Patient is unable to hold urine (incontinent).	
MJW	Patient is unable to pass urine.	
MJX	Patient is passing small quantities of urine frequently.	
MJY	Amount of urine passed in 24 hours . . . (indicate number in decilitres: 1 decilitre equals one-sixth of a pint).	
	Urinary functions normal.	MEP
MKA	Urine contains albumen.	
MKB	Urine contains sugar.	
MKC	Urine contains blood.	
MKD	Urine is very dark brown.	
MKE	Urine is offensive and may contain pus.	
MKF	Penis is swollen.	
MKH	Foreskin will not go back to normal position.	
MKI	Patient has swelling of testicles.	

MKJ	Shall I pass a catheter?	
MKK	I have passed a catheter.	
MKL	I am unable to pass a catheter.	

Diseases of the nervous system and mental diseases

MKP	Patient has headache . . . (Table M I).	
MKQ	Headache is throbbing.	
MKR	Headache is very severe.	
MKS	Head cannot be moved forwards to touch chest.	
MKT	Patient cannot feel pinprick . . . (Table M I).	
MKU	Patient is unable to speak properly.	
MKV	Giddiness (vertigo) is present.	
	Patient is paralysed . . . (Table M I).	MCZ
	Patient is conscious.	MCR
	Patient is semi-conscious but can be roused.	MCT
	Patient is unconscious.	MCU
MKW	Pupils are equal in size.	
MKX	Pupils are unequal in size.	
MKY	Pupils do not contract in a bright light.	
MKZ	Patient has no control over his bowels.	
MLA	Patient has fits associated with rigidity of muscles and jerking of limbs—indicate number of fits per 24 hours.	
	Patient has mental symptoms.	MCY
MLB	Patient has delusions.	
MLC	Patient is depressed.	
	Patient is delirious.	MCX
MLD	Patient is uncontrollable.	
	Patient has attempted suicide.	MGE
MLE	Patient has had much alcohol.	

MLF	Patient has delirium tremens.	
MLG	Patient has bedsores . . . (Table M I).	

Diseases of the heart and circulatory system

	Patient is in pain . . . (Table M I).	MDF
MLH	Pain has been present for . . . (indicate number of minutes).	
MLI	Pain in chest is constricting in character.	
MLJ	Pain is behind the breastbone.	
	Pain radiates to . . . (Table M I).	MDP
	Patient has blueness of face.	MIM
MLK	Patient has pallor.	
	The rate of breathing per minute is . . . (number) (in and out being counted as one breath).	MCE
	The pulse is weak.	MCB
	The pulse rate is irregular.	MBY
	The pulse is too weak to count.	MCC
	The pulse is too rapid to count.	MCD
MLL	Breathing is difficult when lying down.	
MLM	Swelling of legs that pits on pressure.	
MLN	Patient has varicose ulcer.	

Infectious and parasitic diseases

MLR	Rash has been present for . . . (indicate number of hours).	
MLS	Rash first appeared on . . . (Table M I).	
MLT	Rash is spreading to . . . (Table M I).	
MLU	Rash is fading.	
MLV	Rash is itchy.	
MLW	Rash is not itchy.	
MLX	Rash looks like general redness.	
MLY	Rash looks like blotches.	

MLZ	Rash looks like small blisters containing clear fluid.	
MMA	Rash looks like larger blisters containing pus.	
MMB	Rash is weeping (oozing).	
MMC	Rash looks like weals.	
MMD	Rash consists of rose-coloured spots that do not blench on pressure.	
MME	Skin is yellow.	
	Patient has an abscess . . . (Table M I).	MFC
MMF	Patient has buboes . . . (Table M I).	
MMJ	Patient has been isolated.	
MMK	Should patient be isolated?	
MML	I have had (indicate number) similar cases.	
	Patient has diarrhoea with frequent stools like rice water.	MIP
	Patient has never been successfully vaccinated against smallpox.	MUT
	Patient was last vaccinated . . . (date indicated).	MUU
	Patient has vaccination marks.	MUV

Venereal diseases (see also Diseases of genito-urinary system)

MMP	Patient has discharge from penis.	
MMQ	Patient has previous history of gonorrhoea.	
MMR	Patient has single hard sores on penis.	
MMS	Patient has multiple sore on penis.	
	Patient has buboes . . . (Table M I).	MMF
MMT	Patient has swollen glands in the groin.	
MMU	End of penis is inflamed and swollen.	

Diseases of the ear

	Patient is in pain . . . (Table M I).	MDF
MMW	Patient has boil in ear(s).	
MMX	Patient has discharge of blood from ear(s).	

MMY	Patient has discharge of clear fluid from ear(s).	
MMZ	Patient has discharge of pus from ear(s).	
MNA	Patient has hearing impaired.	
MNB	Patient has foreign body in ear.	
	Giddiness (vertigo) is present	MKV
MNC	Patient has constant noises in ear(s).	

Diseases of the eye

	Patient is in pain . . . (Table M I).	MDF
MNG	Patient has inflammation of eye(s)	
MNH	Patient has discharge from eye(s).	
MNI	Patient has foreign body embedded in the pupil area of the eye.	
MNJ	Eyelids are swollen.	
MNK	Patient can not open eyes (raise eyelids).	
MNL	Patient has foreign body embedded in the white of the eye.	
MNM	Patient has double vision when looking at objects with both eyes open.	
MNN	Patient has sudden blindness in one eye.	
MNO	Patient has sudden blindness in both eyes.	
	Pupils are equal in size.	MKW
	Pupils are unequal in size.	MKX
	Pupils do not contract in a bright light.	MKY
	Patient has a penetrating wound . . . , (Table M I).	MFM
MNP	Eye-ball is yellow in colour.	

Diseases of the Skin

See Infectious and Parasitic Diseases (page 113).

Diseases of muscles and joints

MNT	Patient has pain in muscles of . . . (Table M I).	

MNU Patient has pain in joint(s) ... (Table M I).

MNV Patient has redness and swelling of joint(s) (Table M I).

MNW There is history of recent injury.

MNX There is no history of injury.

Miscellaneous illnesses

MLE Patient has had much alcohol.

MOA Patient is suffering from heat exhaustion.

MOB Patient is suffering from heat stroke.

MOC Patient is suffering from sea sickness.

MOD Patient is suffering from exposure in lifeboat—indicate length of exposure (number) hours.

MOE Patient is suffering from frostbite ... (Table M I).

MOF Patient has been exposed to radioactive hazard.

Childbirth

MOK I have a patient in childbirth aged ... (number) years.

MOL Patient states she has had ... (number) children.

MOM Patient states child is due in ... (number) weeks.

MON Pains began ... (number) hours ago.

MOO Pains are feeble and produce no effect.

MOP Pains are strong and effective.

MOQ Pains are occurring every ... (number) minutes.

MOR The bag of membranes broke ... (number) hours ago.

MOS There is severe bleeding from the womb.

MOT The head is coming first.

MOU The buttocks are coming first.

MOV A foot has appeared first.

MOW An arm has appeared first.

MOX The child has been born.

MOY   The child will not breathe.

MOZ   The placenta has been passed.

MPA   The placenta has not been passed.

MPB   I have a non-pregnant woman who is bleeding from the womb.

Progress report

MPE   I am carrying out prescribed instructions.

MPF   Patient is improving.

MPG   Patient is not improving.

MPH   Patient is relieved of pain.

MPI   Patient still has pain.

MPJ   Patient is restless.

MPK   Patient is calm.

MPL   Symptoms have cleared.

MPM   Symptoms have not cleared.

MPN   Symptoms have increased.

MPO   Symptoms have decreased.

MPP   Treatment has been effective.

MPQ   Treatment has been ineffective.

MPR   Patient has died.

Part II Medical advice

Request for additional information

MQB I cannot understand your signal; please use standard method of case description.

MQC Please answer the following question(s).

Diagnosis

MQE My probable diagnosis is . . . (Table M II).

MQF My alternative diagnosis is . . . (Table M II).

MQG My probable diagnosis is infection or inflammation . . . (Table M I).

MQH My probable diagnosis is perforation of . . . (Table M I).

MQI My probable diagnosis is tumour of . . . (Table M I).

MQJ My probable diagnosis is obstruction of . . . (Table M I).

MQK My probable diagnosis is haemorrhage of . . . (Table M I).

MQL My probable diagnosis is foreign body in . . . (Table M I).

MQM My probable diagnosis is fracture of . . . (Table M I).

MQN My probable diagnosis is dislocation of . . . (Table M I).

MQO My probable diagnosis is sprain of . . . (Table M I).

MQP I cannot make a diagnosis.

MQT Your diagnosis is probably right.

MQU I am not sure about your diagnosis.

Special treatment

MRI You should refer to your International Medical Guide for Ships if available or its equivalent.

MRJ You should follow treatment in your own medical guide.

MRK You should follow the instructions for this procedure outlined in your own medical guide.

MRL Commence artificial respiration immediately.

MRM Pass catheter into bladder.

MRN Pass catheter again after . . . (number) hours.

MRO Pass catheter and retain it in bladder.

MRP	Apply ice cold compress and renew every . . . (number) hours.
MRQ	Apply hot compress and renew every . . . (number) hours.
MRR	Apply hot water bottle to . . . (Table M I).
MRS	Insert ear drops . . . (number) times daily.
MRT	Insert antiseptic eye drops . . . (number) times daily.
MRU	Insert anaesthetic eye drops . . . (number) times daily.
MRV	Bathe eye frequently with hot water.
MRW	Give frequent gargles one teaspoonful of salt in a tumblerful of water.
MRX	Give enema.
MRY	Do not give enema or laxative.
MRZ	Was the result of the enema satisfactory?
MSA	Give rectal saline slowly to replace fluid loss.
MSB	Give subcutaneous saline to replace fluid loss.
MSC	Apply well-padded splint(s) to immobilize limb. Watch circulation by inspection of colour of fingers or toes.
MSD	Apply cotton wool to armpit and bandage arm to side.
MSF	Apply a sling and/or rest the part.
MSG	Give light movements and massage daily.
MSJ	Place patient in hot bath.
MSK	To induce sleep give two sedative tablets.
MSL	Reduce temperature of patient as indicated in general nursing chapter.
MSM	The swelling should be incised and drained.
MSN	Dress wound with sterile gauze, cotton wool and bandage.
MSO	Dress wound with sterile gauze, cotton wool and apply well-padded splint.
MSP	Apply burn and wound dressing and bandage lightly.
MSQ	Dress wound and bring edges together with adhesive plaster.
MSR	The wound should be stitched.

MST	The wound should not be stitched.
MSU	Stop bleeding by applying more cotton wool, firm bandaging and elevation of the limb.
MSV	Stop bleeding by manual pressure.
MSW	Apply tourniquet for not more than fifteen minutes.
MSX	Induce vomiting by giving an emetic.
MSY	You should pass a stomach tube.
MSZ	Do not try to empty stomach by any method.

Treatment by medicaments

Prescribing

MTD	You should give . . . (Table M III).
MTE	You must not give . . . (Table M III).

Method of administration and dose

MTF	You should give one tablespoonful (15 ml or ½ oz.)
MTG	You should give one dessertspoonful (7·5 ml or ¼ oz).
MTH	You should give one teaspoonful (4 ml or 1 drachm).
MTI	You should give by mouth . . . (number) tablets/capsules.
MTJ	You should give a tumblerful of water with each dose.
MTK	You should give by intramuscular injection . . . (number) milligrammes.
MTL	You should give by subcutaneous injection . . . (number) milligrammes.
MTM	You should give by intramuscular injection . . . (number) ampoule(s).
MTN	You should give by subcutaneous injection . . . (number) ampoule(s).

Frequency of dose

MTO	You should give once only.
MTP	You should repeat after . . . (number) hours.

MTQ You should repeat every . . . (number) hours.

MTR You should continue for . . . (number) hours.

Frequency of external application

MTT You should apply once only.

MTU You should apply every . . . (number) hours.

MTV You should cease to apply.

MTW You should apply for . . . (number) minutes.

Diet

MUA Give nothing by mouth.

MUB Give water very freely.

MUC Give water only in small quantities.

MUD Give water only as much as possible without causing the patient to vomit.

MUE Give ice to suck.

MUF Give fluid diet, milk, fruit juices, tea, mineral water.

MUG Give light diet such as vegetable soup, steamed fish, stewed fruit, milk puddings or equivalent.

MUH Give normal diet as tolerated.

Childbirth

MUI Has she had previous children?

MUJ How many months pregnant is she?

MUK When did labour pains start?

Give enema.   MRX

MUL Encourage her to rest between pains.

MUM Encourage her to strain down during pains.

MUN What is the frequency of pains (indicate in minutes).

To induce sleep give two sedative tablets.   MSK

MUO	Patient should strain down and you exert steady but gentle pressure on lower part of the abdomen but not on the womb to help expulsion of the placenta.
MUP	You should apply tight wide binder around lower part of abdomen and hips.
MUQ	You should apply artificial respiration gently by mouth technique on infant.

General instructions

MVA	I consider the case is serious and urgent.
MVB	I do not consider the case serious or urgent.
MVC	Put patient to bed lying down at absolute rest.
MVD	Put patient to bed sitting up.
MVE	Raise head of bed.
MVF	Raise foot of bed.
MVG	Keep patient warm.
MVH	Keep patient cool.
MVI	You should continue your local treatment.
MVJ	You should continue your special treatment.
MVK	You should continue giving . . . (Table M III).
MVL	You should suspend your local treatment.
MVM	You should suspend your special treatment.
MVN	You should cease giving . . . (Table M III).
MVO	You should isolate the patient and disinfect his cabin.
MVP	You should land your patient at the earliest opportunity.
MVQ	Patient should be seen by a doctor when next in port.
MVR	I will arrange for hospital admission.
MVS	I think I should come on board and examine the case.

MVT No treatment advised.

MVU Refer back to me in . . . (number) hours or before if patient worsens.

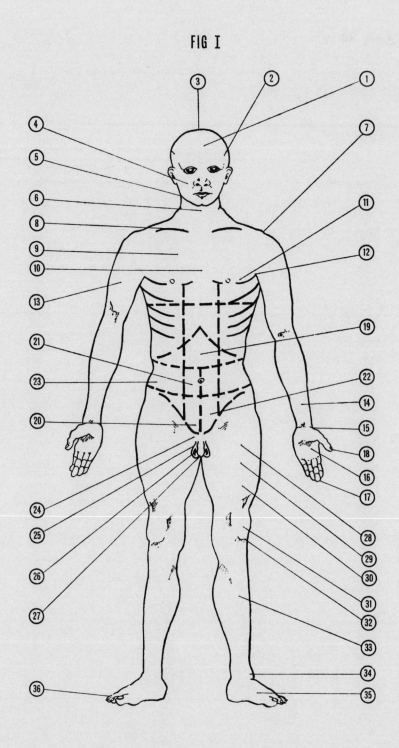

FIG I

Tables of complements

Table M I

Regions of the body Side of body or limb affected should be clearly indicated—right, left.

FIGURE I (Front)

01 Frontal region of head
02 Side of head
03 Top of head
04 Face
05 Jaw
06 Neck front
07 Shoulder
08 Clavicle
*09 Chest
10 Chest mid
11 Heart
12 Armpit
13 Arm upper
14 Forearm
15 Wrist
16 Palm of hand
17 Fingers
18 Thumb
19 Central upper abdomen
20 Central lower abdomen
*21 Upper abdomen
*22 Lower abdomen
*23 Lateral abdomen
*24 Groin
25 Scrotum
26 Testicles
27 Penis
28 Upper thigh
29 Middle thigh
30 Lower thigh
31 Knee
32 Patella
33 Front of leg
34 Ankle
35 Foot
36 Toes

*Indicate side as required.

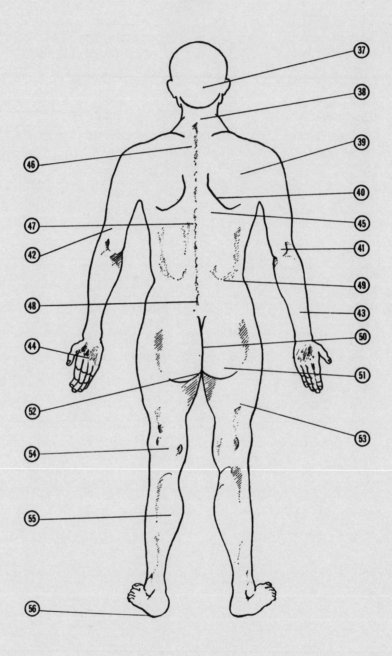

FIG II

FIGURE II (Back)

37 Back of head
38 Back of neck
39 Back of shoulder
40 Scapula region
41 Elbow
42 Back upper arm
43 Back lower arm
44 Back of hand
*45 Lower chest region
46 Spinal column upper

47 Spinal column middle
48 Spinal column lower
*49 Lumbar (kidney) region
50 Sacral region
51 Buttock
52 Anus
53 Back of thigh
54 Back of knee
55 Calf
56 Heel

Other organs of the body

57 Artery
58 Bladder
59 Brain
60 Breast
61 Ear(s)
62 Eye(s)
63 Eyelid(s)
64 Gall bladder
65 Gullet (Oesophagus)
66 Gums
67 Intestine
68 Kidney
69 Lip lower
70 Lip upper
71 Liver
72 Lungs
73 Mouth
74 Nose

75 Pancreas
76 Prostate
77 Rib(s)
78 Spleen
79 Stomach
80 Throat
81 Tongue
82 Tonsils
83 Tooth, teeth
84 Urethra
85 Uterus, womb
86 Vein
87 Voice box (larynx)
88 Whole abdomen
89 Whole arm
90 Whole back
91 Whole chest
92 Whole leg

*Indicate side as required.

Table M II

List of common diseases

01 Abscess
02 Alcoholism
03 Allergic reaction
04 Amoebic dysentery
05 Angina pectoris
06 Anthrax
07 Apoplexy (stroke)
08 Appendicitis
09 Asthma
10 Bacillary dystentery
11 Boils
12 Bronchitis (acute)
13 Bronchitis (chronic)
14 Brucellosis
15 Carbuncle
16 Cellulitis
17 Chancroid
18 Chicken pox
19 Cholera
20 Cirrhosis of the liver
21 Concussion
22 Compression of brain
23 Congestive heart failure
24 Constipation
25 Coronary thrombosis
26 Cystitis (bladder inflammation)
27 Dengue
28 Diabetes
29 Diabetic coma
30 Diphtheria
31 Drug reaction
32 Duodenal ulcer
33 Eczema
34 Erysipelas
35 Fits
36 Gangrene
37 Gastric ulcer
38 Gastro-enteritis
39 Gonorrhoea
40 Gout
41 Heat cramps
42 Heat exhaustion
43 Heat stroke
44 Hepatitis
45 Hernia
46 Hernia (irreducible)
47 Hernia (strangulated)
48 Immersion foot
49 Impetigo
50 Insulin overdose
51 Indigestion
52 Influenza
53 Intestinal obstruction
54 Kidney stone (renal colic)
55 Laryngitis
56 Malaria
57 Measles
58 Meningitis
59 Mental illness
60 Migraine
61 Mumps
62 Orchitis
63 Peritonitis
64 Phlebitis
65 Piles
66 Plague
67 Pleurisy
68 Pneumonia
69 Poisoning (corrosive)
70 Poisoning (non-corrosive)
71 Poisoning (barbiturates)
72 Poisoning (methyl alcohol)
73 Poisoning (gases)
74 Poliomyelitis
75 Prolapsed inter-vertebral disc (slipped disc)
76 Pulmonary tuberculosis
77 Quinsy
78 Rheumatism
79 Rheumatic fever
80 Scarlet fever
81 Sciatica
82 Shingles (herpes zoster)
83 Sinusitis
84 Shock
85 Smallpox
86 Syphylis
87 Tetanus
88 Tonsillitis
89 Typhoid
90 Typhus
91 Urethritis
92 Urticaria (nettle rash)
93 Whooping cough
94 Yellow fever

Table M III
List of medicaments*

A. For external use

01 Auristillae glyceris
   Glycerin ear drops
   Ear Drops

02 Guttae sulfacetamidi
   Sulfacetamide eye drops
   Antiseptic Eye Drops

03 Guttae tetracainae
   Tetracaine eye drops
   Anaesthetic Eye Drops

04 Linimentum methylis
   salicylatis
   Methyl salicylate liniment
   Salicylate Liniment

05 Lotio calaminae
   Calamine Lotion
   Calamine Lotion

06 Lotio cetrimidi
   Cetrimide lotion
   Antiseptic Lotion

07 Naristillae ephedrine
   Norephedrine
   hydrochloride drops
   Nasal Drops

08 Paraffinum molle flavum
   Yellow soft paraffin
   Soft Paraffin

09 Paraffinum molle flavum
   carbasi absorbentis
   Tulle gras dressing
   (paraffin gauze)
   Burn/Wound Dressing

10 Unguentum bacitracini
   Bacitracin ointment
   Antibiotic Ointment

11 Unguentum benzocaini
   compositum
   Compound benzocaine
   ointment
   Pile Ointment

12 Unguentum xylocaini
   hydrochloridi
   Mylocaine ointment
   *Local Anaesthetic
   Ointment*

B. For internal use

Allergic conditions

13 Compressi promethazini
   hydrochloridi
   Promethazine
   hydrochloride tablets
   Antihistamine Tablets
   (25 mg per tablet)

14 Injectic adrenalini
   Adrenaline injection
   Adrenaline
   (1 mg in "ampins")
   Caution: This injection
   No. 14 to be used only on
   medical advice by radio
   except in case of
   anaphylactic shock due to
   penicillin injection

Antibiotics

15 Capsulae tetracyclini
   hydrochloridi
   Tetracycline hydrochloride
   capsules
   Tetracycline Capsules
   (250 mg per capsule)

16 Compressi
   phenoxymethylpenicillini
   Phenoxymethylpenicillin
   Penicillin Tablets
   (125 mg per tablet)

17 Compressi sulfadimidini
   Sulfadimidine tablets
   Sulfonamide Tablets
   (500 mg per tablet)

*Preparations listed above may have been substituted by equivalent preparations in the ship's medicine chest. For the sake of uniformity, medicaments are indicated in the first place by their Latin denomination so that a correct translation can be found in each language.

18 Injectio benzylpenicillini
Procaine penicillin G
Penicillin Injection
(600,000 units per ampoule)

19 Injectio streptomycini sulfatis
Streptomycin sulfate injection
Streptomycin Injection
(1,000 mg per ampoule)

20 Injectio tetracyclini hydrochloridi
Tetracycline hydrochloride
Tetracycline Injection
(100 mg per ampoule)

Asthma

21 Compressi aminophyllini
Aminophylline tablets
Asthma Relief Tablets
(300 mg per tablet)
Caution: This tablet No. 21 to be used only on medical advice by radio

22 Compressi ephedrini hydrochloridi
Ephedrine hydrochloride tablets
Ephedrine Tablets
(30 mg per tablet)

23 Tinctura benzoini composita
Tincture of benzoin compound
Inhalation Mixture

Cough

24 Compressi codeini phosphatis
Codein phosphate tablets
Codein Tablets
(15 mg per tablet)

25 Linctus scillae opiata
Linctus of squill, opiate
Cough Linctus

Diarrhoea

26 Mistura kaolini et morphinae
Kaolin and morphine mixture
Diarrhoea Mixture

Heart

27 Compressi glycerylis trinitratis
Glycerin trinitrate tablets
Heart Tablets
(0·5 mg per tablet)
Note:
For congestive heart failure the following preparations are available on board ship, but they should be used only on medical advice transmitted in plain language and not by code:
Compressi chlorothiazidi (chlorothiazide) or equivalent (500 mg per tablet)
Compressi digoxin (digoxin tablets) or equivalent (0·25 mg per tablet)

Indigestion

28 Compressi magnesii trisilicas
Magnesium trisilicate
Stomach Tablets

Laxatives

29 Compressi colocynthidis et jalapae compositae
Compound colocynth and jalap tablets
Vegetable Laxative Tablets

30 Magnesii hydroxidum
Magnesium hydroxide
mixture
Liquid Laxative—"Milk of Magnesia"

Malaria

31 Compressi chloroquini
sulfatis
Chloroquine sulfate
tablets
Malaria Tablets
(200 mg per tablet)

Pain

32 Compressi acidi
acetylasalicylici
Acetylsalicylic acid
Aspirin Tablets
(300 mg per tablet)

33 Injectio morphini
Morphine sulfate injection
Morphine Injection
(15 mg per ampoule)

Sedation

34 Compressi butobarbitali
Butobarbitone tablets
Sedative Tablets
(100 mg per tablet)

35 Compressi phenobarbitali
Phenobarbitone tablets
Phenobarbitone Tablets
(30 mg per tablet)

36 Compressi chloropromazini
hydrochloridi
Chlorpromazine
hydrochloride tablets
Tranquillizer Tablets
(*Largactil*)
(50 mg per tablet)
Caution: This tablet No. 36
to be used only on
medical advice by radio

Salt depletion or heat cramps

37 Compressi natrii chloridi
solv
Sodium chloride tablets
Salt Tablets
(500 mg per tablet)

Seasickness

38 Compressi hyoscini
hydrobromidi
Hyoscine hydrobromide
tablets
Seasickness Tablets
(0·3 mg per tablet)

Medical index

Note: Numeral references are to pages. Lettered references are to the signals, which are arranged alphabetically in the left hand column on pages 102 to 123.

A

Abdomen, MJG–MJJ

Abscess, MFC

Accidents, MFE–MGT

Admission to hospital, MAC, MVR

Advice
 in childbirth, MUL–MUM, MUO–MUQ
 medical, MQB–MVU
 requested, MAA

Age, MAJ, MAK

Albumen
 in urine, MKA

Alcohol, MLE

Animal bites, MGR

Arm
 immobile, MFY

Arrival, estimated time of, MAD

Artificial respiration, MRL
 on infant, MUQ

Asthma, MIB

B

Bath, MSJ

Bedsores, MLG

Bites, MGR–MGS

Bladder
 catheter: use, MRM–MRO

Bleeding
 diagnosis, MQK
 symptoms, MER, MET
 treatment, MSU–MSW

Blindness, MNN–MNO

Blood
 coughed up, MIF
 discharge from ear, MMX
 in stools, MIQ
 in urine, MKC
 vomited, MIX, MIY

Body regions and organs
 complements tables, 124–127

Boil
 in ear, MMW

Bowels, MEG–MEL, MKZ

Breathing, MCE–MCJ, MLL
 respiratory diseases, MHY–MIB

Bruising, MFQ

Buboes, MMF

Burns, MGG–MGH

C

Capsules, MTI

Carbuncle, MFD

Catheter, MKJ–MKL, MRM–MRO

Childbirth
 answers, MOK–MPB
 medical advice, MUL–MUM, MUO–MUQ
 questions, MUI–MUK, MUN

Chills, MCM

Circulation
 in immobilized limbs, MSC

Circulatory system
 diseases: symptoms, MLH–MLN

Compress
 cold, MRP
 hot, MRQ

Concussion, MFX

Consciousness, MCR–MCV

Constipation, MEJ

Contusion, MFQ

Corrosive, MGJ, MGP

Coughing, MED–MEF, MIC–MIF

Cramp
 stomach, MIU

Cut throat, MGF

D

D.T.s, MLF

Dead, MPR

Delirium, MCX

Delirium tremens, MLF

Delusions, MLB

Depression, MLC

Diagnosis, MQE–MQU

Diarrhoea, MEL, MIP

Diet, MUA–MUH

Digestive system
   diseases: symptoms, MIN–MJP

Diseases
   body area affected, MBE–MBH
   complements tables, p128

Disinfect, MVO

Dislocation, MGA, MQN

Doctors
   instructions for use of Code,
   pp 99–100

Dose
   frequency, MTO–MTR
   quantity MTF–MTN

Double vision, MNM

Dust poisoning, MGQ

E

Ear
   discharging, MMX–MMZ
   diseases, MMW–MNC

Ear drops, MRS

Emetic, MGM–MGO, MSX, MSZ

Enema, MRX–MRZ

Estimated time of arrival, MAD

Exposure in lifeboat, MOD

Eye
   bathing, MRV
   diseases, MNG–MNP
   inflammation, MNG
   pupils, MKW–MKY

Eye-ball
   yellow, MNP

Eye drops
   anaesthetic, MRU
   antiseptic, MRT

Eyelids
   swollen, MNJ

F

Face
   blue, MIM

Fits, MLA

Flatulence, MJD

Fluid loss
   treatment, MSA–MSB

Foreign body
   diagnosis, MQL
   in ear, MNB
   in nose, MGV
   in pupil area of eye, MNI
   in white of eye, MNL
   in wound, MFW
   swallowed, MGL

Fracture, MFY–MGD, MQM

Frostbite, MOE

G

Gangrene, MGT

Gargling, MRW

Gas poisoning, MGQ

Genito-urinary system
   diseases, MJS–MKL
   (*see also* Venereal diseases)

Giddiness, MKV

Glands
   groin, MMT

Gonorrhoea, MMQ

Groin
   glands, MMT

Gums
   diseases: symptoms, MHK

Gun-shot wound, MFV

H

Haemorrhage
  diagnosis, MQK
  symptoms, MER, MET
  treatment, MSU–MSW

Head
  immobile, MKS

Headache, MKP–MKR

Haemorrhoids, MJO–MJP

Hearing
  impaired, MNA

Heart
  diseases, MLH–MLN

Heat exhaustion, MOA

Heat stroke, MOB

Hernia, MJK–MJN

Hiccoughs, MIT

Hoarseness, MHT

Hospital admission, MAC, MVR

Hot bath, MSJ

Hot compress, MRQ

Hot water bottle, MRR

I

Illness
  length, MAM, MAN

Incontinence, MJV

Infection
  diagnosis, MQG

Infectious diseases, MLR–MML

Information
  request for additional, MQB–MQC

Inflammation
  diagnosis, MQG

Injection
  intramuscular, MTK, MTM
  subcutaneous, MTL–MTM

Injuries, MFE–MGT
  body area affected, MBE–MBH
  history, MNW–MNX

Instructions
  for treatment, MVA–MVU
  to users of Code, pp 99–101

International medical guide for ships, MRI–MRK

Intramuscular injection, MTK, MTM

Isolation, MMJ–MMK, MVO

J

Joints
  diseases: symptoms, MNU, MNV

L

Labour pains, MON–MOQ, MUK–MUN

Laxative, MRY

Leg
  immobile, MFZ
  swelling, MLM

Limb immobilization, MSC–MSF

Lips
  swollen, MHA

Localization of symptoms, MBE–MBH

M

Massage, MSG

Masters
  instructions for use of Code, p 99

Medical guide, MRI–MRK

Medicaments
  given, MAU
  prescribing, MTD–MTE

Medicaments, external
  complements tables, p 129
  frequency of application, MTT–MTW

Medicaments, internal
  complements tables, pp 129–131
  dosage, MTF–MTR
  method of administration, MTF–MTN

Mental diseases
  symptoms, MKP–MLF

Mental state, MCR–MDD

Mouth
  diseases: symptoms, MHA–MHK
  ulcer, MHG

Mucus
  in stools, MIR

Muscles
   diseases: symptoms, MNT

N

Nausea, MEO

Nervous system
   diseases: symptoms, MKP–MLF

Nose
   discharge, MGU
   diseases: symptoms, MGU, MGV
   foreign body in, MGV

O

Obstruction
   diagnosis, MQJ

Onset, MBP–MBQ

P

Pain, MDF–MDY
   digestive system, MIU, MJN
   genito-urinary diseases, MJS–MJU
   heart and circulatory system diseases, MLH–MLJ
   in childbirth, MON–MOQ, MUK–MUN
   muscles and joints, MNT–MNU
   nose and throat diseases, MHQ
   respiratory diseases, MHY

Pallor, MLK

Paralysis, MCZ

Parasitic diseases, MLR–MML

Patient
   care of, MVC–MVH
   condition, MAO–MAR; progress report, MPE–MPR
   description, MAJ–MAU; in childbirth, MOK–MPB; method of description, p 99
   isolation, MVO
   medical history, MBA–MBD

Penis, MJT–MJU, MKF–MKH, MMP, MMR–MMS, MMU

Perforation
   diagnosis, MQH

Piles, MJO–MJP

Placenta, MOZ–MPA

Poisoning, MGI–MGK, MGM–MGT
   corrosive, MGJ, MGP
   gas, vapour, dust, MGQ
   non-corrosive, MGI

Poisons, MGI–MGK, MGM–MGT

Port
   distance from, MAD–MAF

Pregnancy
   length, MAL, MUJ

Prescribing, MTD–MTE

Procedural signals
   use, p 2, pp 19–20

Progress report, MPE–MPR

Pulse, MBX–MCD

Pupil (eye), MKW–MKY
   foreign body in, MNI

Pus
   discharge from ear, MMZ
   in urine, MKE

R

Radioactivity
   exposure to, MOF

Rash, MEU–MEV
   infectious and parasitic diseases, MLR–MMD

Rectal saline, MSA

Rendezvous, MAB

Respiration, artificial, MRL
   on infant, MUQ

Respiratory system
   diseases: symptoms, MHY–MIM

Restlessness, MDC, MPJ

Rupture, MJK–MJN

S

Saline
   rectal, MSA
   subcutaneous, MSB

Sea sickness, MOC

Sedatives, MSK

Semi-consciousness, MCT

Ship captain's medical guide, MRI–MRK

Shivering, MCM

Shock, MCW

Medical signals are on pages 102 to 123

Medical index

Skin
   condition, MCO–MCP
   diseases: symptoms, MLR–MMF
   rash, MEU–MEV, MLR–MMD

Sleep, MDD
   inducement with sedatives, MSK

Sleeping pills, MSK

Sling, MSF

Snake bites, MGS

Speech
   incoherent, MKU

Splints
   limb immobilization, MSC

Sprain
   diagnosis, MQO

Sputum, MIG–MIL

Stomach
   cramp, MIU
   emptying, MSX–MSZ

Stomach tube, MSY

Stools, MIN–MIR

Subcutaneous injection, MTL–MTM

Subcutaneous saline, MSB

Sugar
   in urine, MKB

Suicide attempt, MGE

Sunstroke, MOB

Swallowing, MHQ–MHR

Sweating, MCL–MCP

Swelling, MEW–MFD
   incision and drainage, MSM
   penis, MKF
   testicles, MKI
   throat, MHO–MHP
   tongue, MHF

Symptoms, MBP–MPB
   body area affected, MBE–MBH

T

Tablets, MTI

Temperature, MBR–MBW
   reduction, MSL

Testicles, MJU, MKI

Throat
   cut, MGF
   diseases: symptoms, MHL–MHT

Tongue
   diseases: symptoms, MHB–MHG

Tonsils, MHM–MHN

Toothache, MHV

Tourniquet, MFH, MSW

Treatment, MRI–MSZ
   given, MAS–MAU
   instructions, MVI–MVN, MVT
   medicaments, MTD–MTW

Tumour
   diagnosis, MQI

U

Ulcer, MHG–MHJ
   varicose, MLN

Unconsciousness, MCV

Urine, MEP–MEQ
   genito-urinary diseases, MJS, MJT, MJV–MKE

V

Vapour poisoning, MGQ

Varicose ulcer, MLN

Venereal diseases, MMP–MMU

Vertigo, MKV

Voice
   hoarseness, MHI

Vomiting, MEM–MEO, MIU–MJC

W

Wind, MJE–MJF

Womb
   bleeding from, MOS, MPB

Wounds, MFK–MFW
   dressing, MSN–MSQ
   foreign body in, MFW
   treatment, MSN–MST

General index

Note: Numeral references are to pages in chapters 1 to 10 and to the tables of complements. Lettered references are to signals. Single-letter signals are on pages 21 to 25. Two-letter signals are on pages 29 to 94. The signal codes are arranged in alphabetical sequence in the left hand column on those pages.

A

Abandon ship, AC–AI

Accident
 aircraft circling over, BJ
 nuclear, AD, AJ, AK, EC
 oil patches, GC3
 position indicating signals, FC, FD, FF–FK
 proceeding to, FE, SB

Acknowledging, YH, YI

Addressee
 definition, p2
 general instructions, p4

Adrift, CB5, DU–DW, RC

Affirmative, C

Agricultural products
 cargo, SU

Aground, JF–JM
 assistance required, CB4, CI

Ahead, QD, QF–QH

Air temperature, WV

Aircraft
 alighting, AU–BB
 engine trouble, BL
 help in search, BP
 homing, FQ
 hostile, NE7
 in flight, BH–BL, WG
 landing, AU–BB
 magnetic bearings, BZ, CA
 magnetic course, BW–BY
 SAR aircraft, CP1
 sighted, BH
 speed, BQ

Aircraft in distress, DZ–EB
 afloat, BG
 assistance, CM, CR
 communication, EN–EQ
 communication established:
  frequencies, BC–BE
 disabled, F, DR, DS
 ditched, BF, GP
 inability to locate, EO
 missing, GD–GG
 position, EL, EM, EQ, FF–FK, GF, GG
 signals, ED–EK, FC, FD

Alongside, QN–QP, QR

Anchorage, KP, RE–RI, RW, RZ1, ZZ

Anchoring, QS–QY, UB

Anchors
 diver required, IN4
 dragging, Y, RB
 dropping, QN2
 fouling, QT, RA
 letting go, MY7, QZ
 refloating, JT
 weighing, RD

Answers, p1

Approaching
 warnings, MY6

Arms signalling, pp 14–15

Arrival, estimated time of, EQ1, FE1, UR
 aircraft, BP

Assistance
 aircraft help in search, BP
 given, CM, CN, WC1
 not given, CO, CV, WD
 proceeding to, CP–CU, FE
 required, V, W, CB–CJ, HM, NC, TZ, WC
 not required, CK, CL

Astern, S, QI, QK–QM

Astronomical observations, EW2

At the dip
 definition, p2

Atmospheric pressure, WP–WU

Azimuth, p5

B

Ballast, SV

Bar, LP–LS

Barometer, WP–WR

Beaching, JN
 (*see also* Landing)

Beacons, LK, LL

Beacons, radio, EG1–EG8, EW3, FJ4, LM

Bearings, CF, EZ, LT–LW, OX, UK
 general instructions, p5
 magnetic, BZ, CA
 radar, OM, ON
 visual, EW1

Beaufort Scale, XY, YC

Berth, KP, RE–RI

Black frost, VT

Boarding, RS, SQ3

Boats, SN, SQ2
   emergency/distress use, CW–DQ
   motor, DG, DH1, DH3

Bodies
   location, HK
   number, AO2, TB
   picked up, HC

Boiler
   explosion, JD

Boiler room
   damage, IA5
   fire, IV2

Boom, NE3

Bottom
   diver required, IN2

Bottom plate
   damage, IA4

Bottom trawl
   warnings, MY7

Bottom trawling, TE

Bound for, UT, UU

Breeches buoy, GS

Bulkheads, watertight, KE

Bunkers, UZ, VB, VD
   refloating, JP2, JQ2

Buoys, HW5, LK, LL, PA2, PA3, PL, TJ
   breeches buoy, GS

Burns
   doctor required, AN1

C

CO2 fire extinguishers, JA2, JA5

Cable
   chain, LB
   telegraph, RA1
   towing, GV, KS–LG

Calling, CQ, YL, YM

Canal, LX, LY

Cancel, YN, ZP

Carbon dioxide fire extinguishers, JA2, JA5

Carbon tetrachloride fire extinguishers, JA3, JA6

Cargo, ST–SW
   dangerous, B, J, IT, SW
   explosion, JD2
   fire, IV3
   refloat, JP1, JQ1

Carrier of transmitter, OP

Casting off, LG

Casualties, HV–LJ
   doctor required, AN2
   explosion casualties, JE
   number, AP

Caution
   warnings, NE, TH–TJ

Chain cable, LB

Channel, LZ–MC, WE, WG
   swept, OZ, PA

Clearance, health, Q, QQ, ZZ

Close up
   definition, p2

Clouds, VG–VI

Coal, VC2
   cargo, SU1

Coast, PQ
   dangerous, NE1
   minefield, OW1

Coast station
   communication, YP, YU
   distress signals, EJ

Collision, HV–HZ

Collision mat, IN3, KA

Coming from, UT1, UV

Communicating, F, K, YO–ZE

Communications, YH–ZR
   aircraft in distress, EN–EQ;
     frequencies, BC–BE
   vessel in distress, F, EN–EQ, FM, FN

Compartments
   flooded, KC1, KD, IZ1

Compasses, OQ

Complements, p 1
   tables for use with single-letter signals, p 23
   tables for use with two-letter signals, p 9

Single-letter signals are on pages 21 to 25

Two-letter signals are on pages 29 to 94

General index

Consol, EW7

Contamination
  radiation danger, MQ, MR

Convoy, WJ, WK

Co-operation, UW1

Correct, OK

Course, p5
  altering, E, I, MH, MI, MY4, MY5, OJ2, OJ3
  instructions, FL, MG, PI, PM1
  magnetic, BW–BY
  statements, EV, FB, GR, MD–MG, PI1, PI2, PJ
  warnings, MY2, MY4, MY5

Crew, KN2, SZ–TC
  helicopter aid, AZ, BA1, BR1, BT1
  rescue operations, AV, DA, GN, GO, GW, GZ, HA, HD

Crew's quarters
  fire, IV4

Cyclone, VL–VN

D

Dairy products
  cargo, SU2

Damage, HV–IJ
  received, HX, IA, IB
  repairable, IC–IG

Danger
  radiation, AD, AJ, AK, EC, MQ–MX, NG1
  warnings, U, MY–NJ, UL

Dangerous goods, B, J, IT, SW

Dangerous zone, OS2–OS4

Date
  general instructions, p5

Daylight, JR2, JS2, KK, UQ1, US1

Dead
  location, HK
  number, AO2, TB
  picked up, HC

Dead reckoning, EW

Decca navigator, EW5

Deck landing of aircraft/helicopter, AY, BA, BB

Decompression chamber treatment, IQ

Definitions, p2

Depressions, WT, WU

Depth, MP, NK–NN, NP–NS, PW2, QA, QB
  bar, LP, LQ, LR1
  channel/fairway, LZ4, MA

Derelict, GC2, GD2, MJ–ML

Destination
  station of: definition, p2

Diesel oil, VC1

Dip (signal position), p2

Direction-finder, FI, OO–OQ

Disabled vessel/aircraft, F, DR–DT

Discharging oil, CZ1, SX, SY

Distance
  general instructions, p6
  radar, OM, ON

Distress, AA–HT
  assistance required, CB–CJ, NC
  contact or locate, EN–EQ
  position, BI, EL, EM, FC, FD, FF–FK
  signals, ED–EK
  vessel/aircraft, DZ–EC

Disturbance on board, CB3

Ditched aircraft, BF, GP

Diver, A, IN–IR

Doctor, AL–AN
  sent by helicopter, BR2, BT2

Draught, JI1, JJ, JR6, JS6, NT–OG, US5

Drift
  survival craft, FP
  tide, PW1, PZ

Drift nets, TF1, TG1

Drifting
  derelict, ML
  fishing, TE5
  ice, HJ, HO
  vessel, CB5, DU–DW

Dutch, ZAO

E

Echo
  radar, ON

Single-letter signals are on pages 21 to 25

Two-letter signals are on pages 29 to 94

Emergency, AA–HT

Emergency position—indicating radio beacons, EG1–EG8

Enemy
  aircraft, NE7
  submarine, NE6
  vessel, NE5

Engine room
  damage, IA6
  fire, IV1

Engines, IL1, KK4, RJ–RN
  aircraft, BL
  going ahead, QD1, QG4
  going astern, S, QI1, QL4

English, ZA1

Equipment
  cargo, SU4

Escort requested, IM, JV

Estimated time of arrival, EQ1, FE1, UR
  aircraft, BP

Exercise signals, ZF–ZH

Exercises, UY

Explosion, EK1, JB–JE

Extinguishers, fire, JA1–JA6

F

Fairway, LZ–MC, OY2, OY3

Figure-spelling table, p18

Fire, J, IT–JA
  assistance required, CB6

Fire extinguishers, JA1–JA6

Fire-fighting appliances, JA

Firing
  open fire, SN, SQ1

Firing range, NE4

Fishcatch carrier boat, TD

Fishery limits, TW

Fishery protection vessel, TX, TY

Fishing, TE–TG, TS–TV

Fishing gear, NB, TK–TN, TP, TQ, TU1

Fishing vessel, TD, TH

Flag signalling, pp 7–8
  methods, p 3
  procedural signals, p 20

Flags
  aircraft landing aid, AW
  boat landing aid, DC, JN1, RR1

Flame float, FJ1

Flare, FC1, FD1

Flashing light signalling, pp 9–10
  methods, p 3
  procedural signals, p 19

Float
  flame, FJ1
  in tow, KJ1

Floating trawl, TE1

Foam fire extinguishers, JA1, JA4

Fog, XP

Following, WH, WI

French, ZA2

Frequencies, YL, YW, YX
  aircraft in distress, BC–BE
  distress, EG1–EG8

From, DE

Fruit products
  cargo, SU3

Fuel oil, VC

Fumigation, VE

G

Gale, VJ

General call, CQ

German, ZA3

Goods
  dangerous, B, J, IT, SW
  (*see also* Cargo)

Greek, ZA4

Grounding
  assistance required, CB4, CI

Group
  definition, p2

Growlers, VX, VY

H

Hailing distance, PR3

Half water, JK1

Hand-flags signalling, pp 12–15

Harbour, RZ1, UM–UQ

Hatchways, IA7

Hauling nets, G

Hauling warps, TS5–TS9

Hawser, GV, KS–LG, RO1

Headway, QE

Health clearance, Q, QQ, ZZ

Heave to, SQ, UB

Heavy equipment/machinery
    cargo, SU 4

Heel, JW2

Helicopter
    aid/assistance, BR–CA
    landing, AU–BB
    (*see also* Aircraft)

Helicopter in distress *see* Aircraft
    in distress

High water, JK, PW2, QA, QC

Hoist
    definition, p2

Hold
    flooded, KB, KC
    on fire, IV3

Holding ground, RH, RI

Homing vessel/aircraft, FQ

Hostile
    aircraft, NE7
    submarine, NE6
    vessel, NE5

Hurricane, VL–VN

Hydrography, LK–QC

I

Ice, VO–VT
    collision, HW7
    floating, HW7, VZ1
    survivors on drifting ice, HJ, HO

Ice patrol ship, VV

Icebergs, VW–VZ
    collision, HW6

Ice-breaker, VS2–VS7, WC–WF, WL–WO

Ice-field, VU

Identity signals, CS, VF
    allocation, p 4
    definition, p 2
    use, p 4

Infected area, ZV

Inflammable oil
    warnings, MY8

Inflatable rafts, BM3, BR4, BT4

Information, UX

Injured, AO–AT, HA
    helicopter aid, AZ1, BA2, BR3, BT3, BU
    number, TC

Interference
    radar, OH2

International Code of Signals, YU, YV,
    p 1

International Sanitary Regulations,
    ZS–ZZ

International Telecommunication
    Union, p3

Italian, ZA5

J

Japanese, ZA6

Jettisoning, JP, JQ
    warnings, MY8

K

Keep clear, A, D, J, T, IR, RU, UY

L

Landing, DC, DD1, RP, RR
    aircraft/helicopter, AU–BB
    (*see also* Beaching)

Language, YZ–ZC

Latitude
    general instructions, p5

Leak, CB7, JW–JZ

Letter-spelling table, p18

Lifting/picking up operations by helicopter, AZ, BA1, BA2, BR1, BR3, BT1, BT3, BU

Light signal
  marking parachuted object, BM

Lighthouse, CH1

Lights, LN, PG
  aircraft landing aid, AW
  boat landing aid, DC, JN1, RR1
  navigation lights, PD, PE

Lightvessel, CH1, LO
  collision, HW1

Line, GV, KS–LG

Line throwing apparatus, GT, GV1–GV3

Listing, CB1, NI, NJ

Livestock
  cargo, SU5

Local signal codes
  general instructions, p6

Long lines, TF4, TG4
  fishing, TE2

Longitude
  general instructions, p5

Loran, EW6

Lost boat/raft, DM–DQ

Loud hailer, p 3
  procedural signals, p 19

Low water, JK2, PW2, QB, QC1

Lumber
  cargo, SU6

M

MAFOR Code, XV

Machinery
  cargo, SU4

Magnetic bearings, BZ, CA

Magnetic course, BW–BY

Making fast line, LB, LC

Man overboard, O, GW

Manoeuvres, QD–SQ
  intended, X, RT1

Manoeuvring
  trials, RU1
  with difficulty, D, RU

Maritime Declaration of Health, ZT, ZU

Mayday/SOS, ED–EK, FF–FI

Medical assistance, W, HM

Medical Officer, Port, ZW

Medical signals, p 1

Meteorology, VG–YD

Minefield, OW, OX

Mines, OR–OV, OY–PA, PB1, PC, TO

Minesweeping, PB

Missing vessel/aircraft, GD–GG

Mooring, RG

Morse, p 3, pp 14–15, p 17

Mothership, TD1

Motor boat, DG, DH1, DH3

Mutiny, CB3

N

Name
  requested, CS

Names
  vessels/places: general instructions, p4

Navigation, LK–QC
  closed, NA
  dangerous, MZ, VZ
  ice effects, VS, VZ
  instructions, PH–PS, TH–TJ

Navigation buoy
  collision, HW5

Navigation channel, WE

Navigation lights, PD, PE

Negative, N, NO

Nets, G, P, Z, TE–TG, TI, TJ, TO, TS9, TU

No
  visual or sound signal, N
  voice or radio signal, NO

Norwegian, ZA7

Single-letter signals are on pages 21 to 25

Two-letter signals are on pages 29 to 94

Nothing can be done until . . . , US

Nuclear accident, AJ, AK
    abandon ship, AD
    vessel/aircraft in distress, EC

Numbers
    general instructions, p4

Numeral group
    definition, p2

O

Object, submerged, HW4, MO

Obstruction, P, MC, TV1

Oil
    diesel oil, VC1
    discharging, CZ1, SX, SY
    fuel, VC
    inflammable, MY8
    on fire, IV5
    patches, GC3

Oil products
    cargo, SU7

Open channel, WG

Open fire, SN, SQ1

Origin
    station of, p 2
    time of, p 2, p 6

Originator
    definition, p 2
    general instructions, p 4

Overboard, O, GW

Overtaking, PP1

P

Pack ice, VZ2

Pair trawling, T

Parachute, BM, BO

Passenger's quarters
    on fire, IV4

Passing, PO, PP2–PP6

Patrol ship, VV

Persons on board, KN2, SZ–TC
    helicopter aid, AZ, BA1, BR1, BT1
    rescue operations, AV, DA, GN, GO, GW, GZ, HA, HD

Petroleum products
    cargo, SU7

Phonetic tables, pp 18–19

Pilot, G, H, KH, NA2, UA–UE

Pilot boat, UF, UI–UK

Pilotage, radar, OL

Pleasant voyage, UW

Poison
    explosion effects, JD4

Port, KP, KP1, OL, OY, RV2, UH, UL, UM, VB

Port Medical Officer, ZW

Portable radio, DH2

Position, ER–FB
    aircraft in distress, EL, EM, EQ, GF, GG
    distress signals, FC, FD, FF–FK
    homing vessel to, FQ
    survivors, HT
    vessel in distress, BI, CC, EL, EM, EQ
    warnings, MY1

Pratique messages, Q, ZS–ZZ

Pratique signal, ZX

Pressure, atmospheric, WP–WU

Procedural signals, pp 19–20, AA, AB, BN, CQ, CS, DE, NO, OK, RQ, WA, WB
    definition, p 2

Proceed
    orders, GP, HN, RV–RY, UL, WO
    statements, P, CP, CR, FE, GQ, IJ–IL, KA1, SA, SB
    warnings, MY2, MY3, NE

Propellers, CB8, IA9, RO, TP
    diver required, IN1

Pyrotechnic signals, HT

Pumps, water, JA7, JW3

Purse seine, TF, TG

Q

Questions, p 1, RQ, YK

R

Radar, EW4, OH–ON

Radar pilotage, OL

Radiation casualties, AN2

Radiation danger, MQ–MX, NG1
   abandon ship, AD

Radio, SN
   portable radio in boat, DH2

Radio beacons, EG1–EG8, EW3, FJ4, LM

Radio direction-finder, FI, OO–OQ

Radio direction-finder station, EZ

Radio Regulations of International Telecommunication Union, p 3

Radio signals, LW1, WM, ZI–ZO

Radio stations
   calling, CQ
   identification, CS

Radioactive material, MQ, MR, MV–MX

Radiotelegraph installation, DH2

Radiotelegraphy, YW
   procedural signals, p 20
   signalling methods, p 3

Radiotelephony, YX, YY
   general, p 11
   procedural signals, pp 19–20
   signalling methods, p 3

Rafts
   dropped by parachute, BM3
   emergency/distress use, CW, CY1, CY2, CZ–DB, DH, DK, DM–DQ
   helicopter aid, BR4, BT4
   inflatable, BM3, BR4, BT4

Reading, YT

Receiving station
   definition, p 2

Reception, ZI–ZO

Refloating, JO–JV

Repairs to damage, IC–IG

Repeat, AA, AB, BN, WA, WB, ZK, ZN, ZP–ZR

Report on board, P

Report me to . ., ZD1-ZD4

Rescue, GM–GW
   results, GX–HD

River mouth, UQ

Rocket, FC1, FD1, GU

Rocket-line, GT1

Rope, GV, KS–LG, RO1

Routeing of ships, YG

Russian, ZA8

S

SAR aircraft, CP1

SOS/Mayday, ED–EK, FF–FI

Safety signal, YI, YJ

Sanitary Regulations, ZS–ZZ

Scuttling, SN

Sea
   conditions, WW–XB
   marker, FJ2
   marker dye, FJ3
   proceed to sea, P, RV3, UL

Search, FR–GB
   aircraft aid, BP
   for boat/raft, DM–DQ
   information, FM–FP
   instructions, FL, FQ
   results, GC–GL

Searchlight, AX, FC4, FD4, LV, PG2

Seine net, TF2, TG2

Set
   of survival craft, FP
   of tide, PW1, PY

Shallow water, MP

Shelter, KP2

Shoal, MO

Shooting
   open fire, SN, SQ1

Sick, AO–AT, HA
   helicopter aid, AZ1, BA2, BR3, BT3, BU
   number, TC

Side plate
   damage, IA2, IA3

Sighting
   aircraft, BH

Single-letter signals are on pages 21 to 25

Two-letter signals are on pages 29 to 94

Signalling, p 1
　　arms, pp 14–15
　　flags, p 3, pp 7–8, p 20
　　flashing lights, p 3, pp 9–10, p 19
　　hand-flags, pp 12–15
　　methods, p 3
　　procedural, p 2, pp 19–20
　　single-letter, p 1
　　sound, p 2, p 3, p 16
　　three-letter, p 1
　　visual, p 2
　　voice, p 3, pp 18–19

Signals
　　assistance requesting, CF
　　distress, ED–EK, FC, FD, FF–FK
　　exercising, ZF–ZH
　　homing vessel/aircraft, FQ
　　ice-breaker support, WM
　　identity, p 2, p 4
　　light, BM, BM1
　　pratique, ZX
　　procedural, pp 19–20, AA, AB, BN,
　　　　CQ, CS, DE, NO, OK, RQ, WA, WB
　　pyrotechnic, HT
　　radio, LW1, WM, ZI–ZO
　　safety, YI, YJ
　　single-letter, pp 21–25
　　smoke, BM, FC5, FD5, FJ1
　　visual, FC, FD, WM
　　watch for, X, RT, WM

Single-letter signalling, p 1

Single-letter signals, pp 21–25

Sinking, DX, DY, HY

Smoke float, FJ1

Smoke signals, BM, FC5, FD5, FJ1

Sound signalling, p 2, p 3, p 16
　　voice, p 3, pp 18–19

Space ship, CH2

Spanish, ZA9

Speed, A, EV, FB, GR, IL, LH–LJ, RY,
　　SG–SL, WI
　　aircraft/helicopter, BQ
　　general instructions, p 6

Speed trials, SM

Station of destination
　　definition, p 2

Station of origin
　　definition, p 2

Single-letter signals
are on pages 21 to 25

Two-letter signals
are on pages 29 to 94

Stations
　　calling, CQ
　　definition, p 2
　　identification, CS
　　radio, CQ, CS
　　receiving, p 2
　　transmitting, p 2

Steering
　　gear, CB2, IA8
　　instructions, PH–PS

Stem, IA

Stern frame, IA1

Sternway, QJ

Stop, L, M, RT, SN–SQ

Stopping
　　warnings, MY

Storm, VK–VN

Stretcher, BS

Submarine survey work, IR

Submarines
　　collision, HW2
　　exercising, NE2
　　hostile, NE6

Submerged object
　　in tow, KJ
　　struck, MO

Superstructure
　　icing, VT

Surface craft
　　collision, HW

Survival craft
　　assistance required, CN2
　　distress signals sighted, EK2
　　position, FJ, FK, FP, GH, GI, HG

Survivors, CN1, GC2, GJ1, HB, HF–HT

Sweep, TO

Swell, ND, XC–XH

T

Tables
　　complements, p 95
　　figure-spelling, p 18
　　letter-spelling, p 18
　　morse, p 17
　　phonetic, pp 18–19

Tackline, p 2

Tank
　explosion, JD1

Target
　in tow, KJ2

Telegraph cable, RA1

Temperature, WV

Three-letter signals, p1

Tide, JI3, JK, JR3, JS3, PT–QC, US2

Timber
　cargo, SU6
　required, KE1

Time
　general instructions, p 6

Time of origin, p 2, p 6

Towing, KJ–KR

Towing line, GV, KS–LG

Towing line fishing, TE3

Toxic effects
　of explosion, JD4

Transmission, ZI–ZO
　bearings, LW

Transmitter carrier, OP

Transmitting station, p2

Trawl
　bottom, MY7
　floating, TE1

Trawling, T, TE–TG, TS9, TU, TV1

Tropical storm, VL–VN

Tsunami, ND

Tugs, Z, JR7, JS7, KF–KI, NA1, QS2, US6

Two-boat fishing, TE4

Two-letter signals, p1

Typhoon, VL–VN

U

Uncharted obstruction, MC

Under way, SC–SF

Underwater object
　collision, HW4

Underwater operations, IR

Unknown stations
　call signs, CQ

V

VHF radiotelephony, YY

Veering
　of boat/raft, DB
　of lines/warps, LD–LF, TS1–TS4

Vessel
　collision, HW, HY, HZ
　fishing, TD, TH
　homing, FQ
　hostile, NE5
　identification, CS
　in tow, KM–KO, LH
　sunk, HY
　unknown, HW3

Vessel in distress, DZ–EQ
　adrift, CB5, DU–DW
　aircraft assistance, BI, BJ, BP
　assistance, CM, CR
　communication with, EN–EQ, FM, FN
　disabled, F, DR, DT
　drifting, CB5, DU–DW
　missing, GD–GG
　position, BI, CC, EL, EM, EQ, GF, GG
　signals, ED–EK, FC, FD, FF–FK, FQ
　sinking, DX, DY

Visibility, EO, JR4, JS4, US3, XI–XP

Visual bearings, EW1

Visual signal distance, PR2, ZE

Visual signalling, p 2
　arms, pp 14–15
　flag, p 3, pp 7–8, p 20
　flashing lights, p 3, pp 9–10, p 19
　hand flags, pp 12–15

Visual signals, FC, FD, WM

Voice signalling, p 3, pp 18–19

Voyage, pleasant, UW

W

Wake, PM, UG

Warnings, MY–NJ

Warps, TS, TU

General index

Water
    depth, LP, LQ, LR1, LZ4, MA, MP, NK–NN, NP–NS, PW2, QA, QB
    half, JK1
    high, JK, PW2, QA, QC
    leak, CB7, JW–KE
    low, JK2, PW2, QB, QC1
    shallow, MP

Water-line
    damage, HX1–HX4

Water pumps, JA7, JW3

Watertight bulk heads, KE

Waves, ND, XC–XH

Way, M, SP

Weather, VG–YD
    moderating, US4, XU
    permitting vessel to get under way, SC2
    rescue conditions, GQ
    state, XQ, XR, XT
    warnings, MY3, ND, XU

Weather forecast, XV
    sea, WZ–XB
    swell, XF–XH
    visibility, XM–XO
    wind, YA–YD

Weather report, XS

Whip, GS

Wind, XW–YD

Wire
    towing, LB2, LB3

Wire hawser, KU2, KW5, KW6

Wireless, SN

Wreck, MM, MN

Wreckage, FJ5, GD2, GJ–GL, GY, MK1

Y

Yes, C